The Discovery of Weather

The
Discovery
of Weather

Stephen Saxby, the tumultuous birth of weather forecasting, and Saxby's Gale of 1869

Jerry Lockett

Formac Publishing Company Limited
Halifax

Formac Publishing Company Limited recognizes the support of the Province of Nova Scotia through the Department of Communities, Culture and Heritage. We are pleased to work in partnership with the Culture Division to develop and promote our culture resources for all Nova Scotians. We acknowledge the support of the Canada Council for the Arts which last year invested $24.3 million in writing and publishing throughout Canada. We acknowledge the financial support of the Government of Canada through the Canada Book Fund for our publishing activities.

Cover image: iStock

Library and Archives Canada Cataloguing in Publication

Lockett, Jerry
       The discovery of weather : Stephen Saxby, the tumultuous birth of weather forecasting, and Saxby's gale of 1869 / Jerry Lockett.

Includes bibliographical references and index.
Issued also in electronic formats.
ISBN 978-1-4595-0080-8

       1.Weather forecasting--History--19th century. 2.Saxby, Stephen, 1804-1883. 3.Meteorologists--England--Biography. 4.Weather forecasting--England--History--19th century. 5. Windstorms--Atlantic Coast (U.S.)--Forecasting--History--19th century. 6. Windstorms--Atlantic Coast (Canada)--Forecasting--History--19th century. I. Title.

QC995.L62 2012   551.6309   C2012-903524-6

Formac Publishing Company Limited
5502 Atlantic Street
Halifax, Nova Scotia, Canada
B3H 1G4
www.formac.ca

Printed and bound in Canada.

It is the very error of the moon;
She comes more nearer earth than she was wont,
And makes men mad.

—Shakespeare, *Othello*

Contents

Cast of Characters   10

Chronology   13

Introduction   15

Part One—Mr. Saxby's Prediction

Chapter 1: The Calm before the Storm   23

Chapter 2: A Storm of Ideas   49

Chapter 3: Storm Warnings   73

Chapter 4: Barking at the Moon   100

Part Two—The Storm

Chapter 5: Deluge   135

Chapter 6: Landfall   157

Chapter 7: Storm Surge   182

Chapter 8: Aftermath   198

Chapter 9: Why Saxby Still Matters   219

Acknowledgements   244

Appendix 1: The Saffir-Simpson Scale of Hurricane Intensity   246

Appendix 2: The Weather during Saxby's Storm   248

Notes   250

Glossary   262

Bibliography   265

Image Credits   269

Index   270

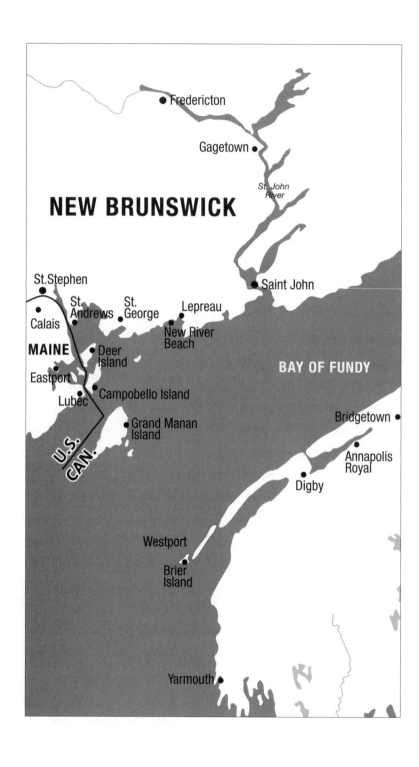

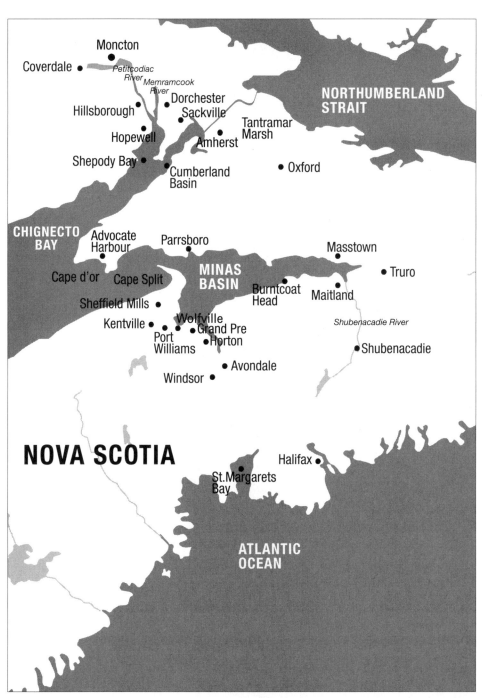

The Bay of Fundy with locations mentioned in the text

Cast of Characters

Cleveland Abbe: American meteorologist, that country's first official weather forecaster.

Sir George Airy: Britain's Astronomer Royal from 1835 to 1881, a supporter of FitzRoy.

Frederick Allison: Amateur meteorologist in Halifax, Nova Scotia.

R. H. Allnatt: Regular London *Times* weather correspondent, and critic of Saxby's lunar theory.

François Arago: French astronomer and physicist, a supporter of Espy in the storm controversy.

Alexander Dallas Bache: American scientist who, as superintendent of the U.S. Coast Survey, became involved in a turf war with Matthew Maury.

Christoph Buys Ballot: Dutch meteorologist and proponent of international cooperation in meteorology.

Clement Horton Belcher: Almanac publisher in Halifax, Nova Scotia.

Gaspard-Gustave de Coriolis: French scientist who discovered the Coriolis effect.

William Cunnabell: Almanac publisher in Halifax, Nova Scotia.

James Pollard Espy: Brilliant but argumentative American scientist, a.k.a. "the Storm King," who discovered that latent heat is the main source of the energy of hurricanes and challenged the views of William Redfield.

William Ferrel: American mathematician who demonstrated that the Coriolis effect affects the motion of winds.

Robert FitzRoy: British Admiral who invented the term "forecast," established storm warnings in Britain, and issued the world's first weather forecast.

Benjamin Franklin: American polymath, and a founding father of the United States, who was the first to realize that storms move from place to place.

James Glaisher: English meteorologist employed at the Royal Observatory at Greenwich.

Robert Hare: American scientist who played a minor role in the great storm controversy.

Joseph Henry: Leading American physicist who, as first secretary of the Smithsonian Institution, became a leading light in the development of weather forecasting in the United States.

Sir John Herschel: English astronomer, son of Sir William Herschel, who took up Redfield's ideas on storms in Britain.

Luke Howard: English amateur meteorologist who classified clouds, and who also believed that the moon influenced weather.

George Templeman Kingston: Pioneer of weather forecasting in Canada, where he initiated storm warnings.

Charles Knight: Publisher of the *British Almanac*.

George Cargill Leighton: Publisher of the *Illustrated London Almanack*.

Urbain Le Verrier: French mathematician and astronomer who foresaw the possibility of storm warnings.

Matthew Fontaine Maury: American oceanographer who brought meteorologists together to agree on an international system for recording weather observations.

Richard James Morrison: Notorious astrologer and almanac publisher in England, a.k.a. Zadkiel Tao-Sze.

Samuel Morse: Inventor of the electric telegraph, without which weather forecasting would never have been possible.

Patrick Murphy: English almanac publisher.

Denison Olmsted: American professor at Yale College who championed the work of William Redfield.

Henry Piddington: English sea captain who developed a rule of thumb that helped sailors avoid the worst winds in a typhoon or hurricane.

William C. Redfield: Amateur American scientist who discovered that the winds in hurricanes rotate *around* a storm centre. Espy's main opponent in the great storm controversy.

William Reid: British army officer who studied Atlantic hurricanes and became a great ally of Redfield.

Stephen Martin Saxby: English naval instructor who believed that the moon's motion influences weather. "Predicted" the hurricane of October 4, 1869, that hit the Bay of Fundy.

Zadkiel Tao-Sze: *See* Richard James Morrison.

Chronology

1743 Benjamin Franklin discovers that storms move from place to place.

1821 William C. Redfield observes that winds in a hurricane rotate around its centre.

1833 James Pollard Espy discovers that latent heat is the source of energy in a hurricane.

1834–59 Redfield and Espy go head-to-head in the American Storm Controversy.

1835 Gaspard-Gustave de Coriolis discovers that moving objects on the surface of the earth are deflected to the right in the northern hemisphere.

1843 Robert FitzRoy first proposes the idea of storm warnings in Britain.

1847 Samuel Morse patents his electric telegraph.

1850 Joseph Henry starts posting a daily weather map at Washington's Smithsonian Institution.

1853 (September) Matthew Fontaine Maury organizes the Brussels Conference on Meteorology.

1853 (November 13) A devastating storm in the Crimea leads many to believe that the telegraph would make storm warnings possible.

1854 (August) Robert FitzRoy becomes head of the British Board of Trade's Meteorological Department.

1856 William Ferrel shows that winds on earth are deflected because of the Coriolis effect.

1861 (February 6) FitzRoy issues his first storm warning for British mariners.

1861 (August 1) FitzRoy's—and the world's—first weather forecast is published in the London *Times.*

1861–65 The Civil War interrupts weather forecasting in the U.S.

1865 Following FitzRoy's death, the British Board of Trade halts weather forecasts.

1868 (December) Stephen Saxby warns of a devastating storm for the following October.

1869 (September 1) Cleveland Abbe issues the first of his weather "probabilities" in the U.S.

1869 (October 4–5) Saxby's Storm.

Introduction

In any year, two, three—sometimes four—Atlantic hurricanes reach the shores or offshore waters of Nova Scotia, New Brunswick, and Newfoundland. Usually they have lost much of their vigour by the time they reach these latitudes. But not always.

The hurricane that hit the Bay of Fundy on October 4, 1869, and became known in the Canadian Maritimes as the Saxby Gale—a bit of an understatement, given its malevolence—was not the most destructive storm in the history of the region. Other historic storms have taken more lives, sunk more ships, and destroyed more buildings. One of the deadliest Atlantic hurricanes, for instance, struck Newfoundland in 1775, killing as many as four thousand people in what was likely the worst natural disaster in Canadian history. Another killer storm, packing winds of a hundred knots (185 km/h), passed just south of Nova Scotia in August 1873, killing at least 223 (and maybe as many as six hundred) people and destroying twelve hundred boats and nine hundred buildings.

Saxby's Gale took a heavy toll, nevertheless. As hurricanes go, it was modest in size, but the enormous power locked up in the thunderheads of a hurricane of *any* size make it a brute to be reckoned with. It has been said that if all the power in a fully developed Atlantic hurricane could be harnessed, it would supply the world's demand for electricity for a month. To imagine how vicious a storm Saxby's Gale was, Nova Scotians need only think back to Hurricane Juan of September 2003—a storm of about the same intensity—which is still fresh in the memories of many for the seven deaths and widespread destruction it caused.

The Saxby Gale was a most unusual storm for a number of reasons.

First, it was "predicted" almost a year ahead by the Royal Navy man for whom it was named. Stephen Martin Saxby was a controversial character whose ideas about weather went against the grain of mainstream meteorology of the time. In many ways he was a typical middle-class Victorian Englishman who liked to dabble in scientific matters, from archaeology to zoology. He was intelligent, well-educated, a humanitarian, and a family man, and he had the courage of his convictions, even when the scientific establishment scorned his ideas. He is often referred to as Lieutenant or Captain Saxby, but he was not in fact a naval officer at all, nor did he ever refer to himself as such. His naval ranks were invented by the newspapers. Saxby was a naval instructor, a civilian who was hired by the navy under rather unusual circumstances.

Second, it caused worse flooding in the eastern United States than had ever been seen before. It dumped record levels of rain on Virginia, Maryland, Pennsylvania, New Jersey, New York, Connecticut, Massachusetts, Vermont, New Hampshire, and Maine, which all measured rainfall of five inches or more. In New England it became known as the Great Northeastern Rainstorm and Flood of 1869.

Third, by all rights the storm should have been severely weakened as it reached northern waters. Instead it met with another weather system near Cape Cod, the two merged, and the resulting storm became explosive. It picked up strength, with winds of 90 knots (166 km/h), then barrelled ashore in the Bay of Fundy just as an extreme tide was working its way up the bay. The resulting storm surge breached defensive dykes that had withstood storms for more than two centuries.

Weather forecasts are such a part of everyday life now that it is hard to imagine just how much the early practitioners of this desperately needed public service struggled for recognition and acceptance. Until the nineteenth century, there was a common belief that storms were born in one place, did their dirty work with their winds blowing in a haphazard way, and then fizzled out. No one realized that they are highly organized creatures, that they move from place to place—and that they can usually be predicted a day or two ahead.

That all changed during Stephen Saxby's lifetime, as knowledge of storms increased sharply. But the whole business of meteorology was in turmoil on both sides of the Atlantic. In the United States, a bitter dispute between scientists about the nature of storms lasted for decades. The men who struggled to initiate forecasting networks became engaged in other disputes and in turf wars, and the Civil War brought the whole enterprise to a grinding halt, from which it took several years to recover.

In Britain, the pioneers of weather forecasting faced severe opposition from a scientific community that condemned their efforts as unscientific. They also faced intense competition for the public's attention from weather prophets with a vested interest— the astrologers and publishers of almanacs, with their outrageous pretense that they could predict the weather months, even a year or more in advance. Finally, after an unexpected tragedy, the British government shut down the country's one forecasting service that *had* started, and that remained dormant for fourteen years.

Meanwhile, Stephen Saxby thought he had found a link between weather disturbances and certain stages of the moon's orbit of the earth. He used his "Lunar Theory" to make long-range predictions of weather that mariners could take to sea with them, confident in the knowledge of when they should be wary of approaching storms without the need for keeping a close eye on shore-based storm signals. How marvellous had he been right, but he was not. Few believed his claims, and ironically, the storm that bears his name made more skeptics than converts to his theory. It's also ironic that, though the storm's existence had nothing whatever to do with the motions of the moon, it was the powerful effect of the moon on the earth's oceans—in generating tides—that turned the storm into the destructive monster it became.

Saxby never achieved the recognition that he hoped for, and he remains largely unknown. Records of his ten-year naval career are preserved in Royal Navy archives, and his books and articles, now long forgotten, reveal further clues about the man and his ideas, but, were it not for that 1869 hurricane, even his name would have faded into obscurity.

The Saxby Gale has inspired a number of things: a pub in Moncton, known as *Saxby's Pub and Eatery*, and a band called the *Saxby Gale*. Patrons can enjoy a rum-based cocktail called the *Saxby Gale* at Toronto's Harbord Room Restaurant & Bar. And at least one child was christened *Saxby* in the aftermath of the storm.

Saxby's Gale also inspired this book.

Part one tells the story of how a handful of scientists in the early part of the nineteenth century began to come to grips with the nature of hurricanes and other major storms. It follows the trials and tribulations of those who tried to forecast the weather by using the scientific knowledge available at the time, although the odds seemed stacked against them. It also recounts the events in the life of Stephen Saxby that led to his unorthodox ideas on weather and to his fortuitous prediction.

Part two follows the storm Saxby "predicted," and the toll it took in the eastern U.S. and in the Canadian Maritimes. It looks into the likely origins and follows the violent passage of this hurricane, from its birth somewhere in the vicinity of the Caribbean and its track northward along the eastern seaboard, to its disastrous landfall at the mouth of the Bay of Fundy where it generated the famous storm surge, the like of which has never yet been repeated. Finally it looks to the future, at the need for people to take seriously the prospect that a storm surge of this magnitude will almost certainly happen again.

PART ONE
Mr. Saxby's Prediction

I saw the new moon late yestreen
Wi' the auld moon in her arm;
And if we gang to sea master,
I fear we'll come to harm.
                    —Eighteenth-century English ballad
                              "Sir Patrick Spens"

CHAPTER 1

The Calm before the Storm

It is bad enough when it comes, without our having the misery of knowing about it beforehand.

—Jerome K. Jerome

On the morning of Sunday June 11, 1815, a throng of sightseers lined the chalk-white cliffs that lie to the east and west of the seaside town of Ramsgate in England. There was always much to see on this busy stretch of coastline, including the merchant ships that plied their trade up and down the English Channel and into the North Sea, bound for the major ports of England and Europe—London, Liverpool, and Rotterdam. There were smaller coastal traders too, fishing fleets, naval transports, and the huge naval men-o'-war that often anchored in the Downs—the anchorage that lies between Kent and the notorious Goodwin Sands in the English Channel—after voyages to the far-flung regions of the British Empire. It was particularly busy right now, owing to the war with Napoleon that within a month would be brought to a conclusion at Waterloo.

But on this particular day the crowds were drawn to the shore by word that had spread like wildfire among the seaside towns of southern England, word of a ship like none they had ever seen. It was indeed a most peculiar-looking vessel. Stocky and broad-beamed, it carried only one large, square sail, hoisted on what looked like a rather fat but squat mast, stepped amidships. Smoke, steam, and an occasional fiery blast of red-hot sparks belched from the peak of this mast, leaving a trailing black cloud hovering in the air to mark the ship's passage. And on each side it carried a large, bladed wheel, almost ten feet in diameter, most of it enclosed above-water within a great, semicircular housing. The wheels rotated steadily, churning the ship's wake into a lingering, frothy whiteness.

The vessel was one of Europe's first sea-going steam-packets, the paddle-driven *Thames*, and it was on the final leg of an astonishing 1,500-mile voyage from Glasgow in Scotland, around Land's End, bound for Limehouse in London. It was one of the first voyages ever attempted by a steamboat on the open sea, and certainly the longest to date by far. The ship had been built on the River Clyde in 1813, and measured seventy-nine feet in the keel. It was powered by one of the earliest steam engines ever used to drive a ship, and although only rated at fourteen horsepower the engine could propel the *Thames* along at a decent clip (it could average almost seven knots) in ideal conditions. The ship had already earned its living for a year, plying between Glasgow and Greenock, but had recently been purchased by a company in London, which planned to operate a twice-weekly return service taking passengers between Wool Quay in the capital and Margate in Kent.

The voyage had been full of incident. Soon after setting out, the *Thames* had been nearly shipwrecked off Portpatrick in Scotland. On at least two occasions during the voyage, observers who saw the dense black smoke emerging from its single funnel thought the ship was on fire and rushed to its assistance. Stormy weather and rough water in the Irish Sea had damaged blades on both paddles, and the blades had had to be cut away. Even so, the *Thames* had shown that in light winds it could outpace a naval sailing ship, the sloop-of-war HMS *Myrtle*, much to the embarrassment of its officers. After rounding Land's End the little packet had been tossed around by tremendous Atlantic swells, but it had survived the ordeal and continued its passage up the English Channel.

The *Thames* created a sensation wherever it docked along the way and attracted crowds of sightseers who had never seen a steam vessel before. At Plymouth, where it arrived on June 6, sailors had swarmed up the rigging of their ships for a view of this novel vessel as it showed off its manoeuvrability among the men-o'-war at anchor in Plymouth Sound. At Portsmouth it had drawn a crowd numbering in the tens of thousands, and had been swarmed by so many small vessels trying to take a close look that the Port Admiral had to assign a guard to the ship. Even a court-martial that had happened to be in progress aboard the frigate *Gladiator* was temporarily halted so that the assembled court could take time to see the vessel.

Among the crowd that trudged its way up Ramsgate's cliffs to watch the *Thames* steam through the Downs that Sunday morning was a ten-year-old boy who was fascinated by all things nautical. No doubt he was familiar with many of the sailing vessels that passed this way, but this odd-looking little ship stirred his imagination.

The boy's name was Stephen Martin Saxby, and the sight of the *Thames* would have a profound impact on his future. Perhaps he already knew of George Stephenson's first steam locomotive, which had been used to haul coal in Wales just a year earlier. And perhaps he had seen the stationary engines that were now beginning to power local factories. Maybe, like many others, he thought what he was seeing was just a one-minute wonder. Or maybe he knew right away that he was looking at the new face of shipping. If that was so, he was way ahead of his time, for few people at the beginning of the nineteenth century dared imagine that steam could ever replace sail. But whatever Stephen Saxby was thinking that day, the experience left its impression on him, an impression that would determine the course of his life and career. Steam engines and steamships were to figure large in his future.

The steam engine—a heat engine—had been around for several years by 1815, although the notion that heat could be transformed into motion was still fairly new. But another type of heat engine, of which Saxby would know nothing, would also play a significant part in his life. This heat engine occurs in nature and is found in the earth's atmosphere, where the heat of the sun sucks warm moisture from the surface of the oceans into the atmosphere. When conditions are just right, this warm moist air rises to create a moving phenomenon millions of times more powerful than even the largest man-made machine. That motion is the enormous and fearsome wind of a tropical hurricane, and it was to an Atlantic hurricane that barrelled up the eastern seaboard of the U.S. and into the Bay of Fundy that the young boy would give his name more than half a century later—the Saxby Gale of 1869.

~~~

Some six decades before Stephen Saxby was born, a gifted Philadelphia scientist made an observation that is crucial to our modern understanding of storms. The man was an inventor, author, publisher, printer, and politician who also had a passion for flying kites and managed to squeeze the roles of diplomat, statesman, postmaster, musician, and much else besides into his busy life. Benjamin Franklin was not only decidedly brainy, but he also crammed so much into his eighty-four years that you wonder how he found time to eat, sleep, or even fly his famous kites.

Franklin took a keen interest in the weather and in astronomy, and he studied both subjects assiduously. In the year 1743, he had been hoping to observe an eclipse of the moon on the night of October 21, but his plans were disrupted by a powerful storm packing ferocious northeast winds. While ordinary mortals fretted about mundane concerns—like roofs becoming airborne and chimneys collapsing—Franklin was miffed because the night sky over the city was obscured by heavy cloud and rain. "Neither the Moon nor the stars could be seen," he later complained to his friend Jared Eliot.

He was quite surprised to later read in the Boston newspapers, the *Boston Evening Post* among them, that the eclipse had been clearly visible in that city, some three hundred miles to the northeast of Philadelphia. Like everyone else at the time, Franklin had assumed that as the winds had blown from the northeast the storm must have travelled to Philadelphia from that direction, and that the storm must have unleashed its fury on Boston first. Anyone in Boston hoping to observe the eclipse, Franklin surmised, would have been as disappointed as he was. But his brother in Boston informed him that the storm—in fact it was a hurricane—had not arrived in that city *until an hour after the eclipse was over.*

Franklin came to a conclusion that is common knowledge today, but then was as startling as it was goundbreaking. Although it seemed to defy common sense, he realized, the storm had been moving, not from the direction from which the wind was blowing (the northeast), but from the *opposite* direction (the southwest).

Franklin pursued his interest in weather, studying whirlwinds, tornadoes, and waterspouts—and a host of other natural phenomena—throughout the rest of his life. As a scientist he is perhaps best known for his experiments with kites and electricity.

Producing an electrostatic spark from a kite, he established that lightning is caused by electricity in the atmosphere. He put this discovery to practical use with his invention of the lightning rod to deflect the massive electrical discharges from thunderstorms from the roofs of tall buildings to the ground. Or perhaps, as Mark Twain put it, "In order to get a chance to fly his kite on Sunday, he used to hang a key on the string and let on to be fishing for lightning"!

But, despite Franklin's contribution, human understanding of the nature of storms would progress no further for several decades.

∽∽∽

It's a safe bet that the day early humans discovered they could talk, they started talking about the weather. And ever since, one of the more earnest topics of weather conversation has always been: what will it be like tomorrow, next week, or next month? People in the late eighteenth and early nineteenth centuries were no exception, but even though they had only a poor understanding of the natural forces that create weather systems, and many scoffed at the idea that it would ever be possible to truly foretell the weather, there was a tremendous thirst for weather predictions. It was this thirst that, along with other factors like improvements in adult literacy, led to a flourishing and profitable publishing industry—the almanacs.

By the time Queen Victoria ascended the British throne, almanacs were immensely popular. They appealed to people at every level of society, were as diverse as the readerships they catered to, and were in almost every British and American household, rich and poor alike. Published annually, they always included calendars, usually giving details of astronomical events, and articles on topics that would appeal to their target audience: farming

Benjamin Franklin

and gardening, for example, along with local or regional reference information on government, politics, and business. And, of course, they nearly always included gems of weather wisdom.

The more popular almanacs were literature for the masses, like the tabloid press of today. Their proprietors—many of them not the most principled of men—knew that a sure way to boost sales was to appeal to their readers' most banal instincts, and the best-selling almanacs were extremely lucrative. *Old Moore's*, one of the most popular, sold an astonishing 362,449 copies in 1801, making a tidy profit of almost £2,600. But by 1838, four years after the repeal of the hefty *Stamp Acts* duty that had nearly doubled the cover price of a typical almanac, *Old Moore's* circulation topped half a million, and the total number of almanacs sold in Britain that year has been estimated at nearly three times that.

The almanacs came in a variety of flavours. At the upper end of the scale was the *Illustrated London Almanack*, an offshoot of the respected and respectable *Illustrated London News*. It included some genuinely scientific articles, and steered clear of the controversial matter of weather prediction. At the other end were almanacs like *Old Moore's* that catered to a less well-educated, more gullible audience and mixed fact indistinguishably with falsehood. Of all the many frauds perpetrated by the more frivolous almanacs, perhaps the most egregious was their pretense that they could predict weather by as much as a year ahead.

Their weather content was one of these almanacs' strong selling points, even though it was of dubious value, or none at all. The predictions were usually absurd and unscientific, but to an unsophisticated readership they were all that was available. Often the predictions were based on folklore and on old-wives tales like the one that says if it rains on St. Swithin's Day (July 15), it will rain for another forty days. Much of this guff was repeated year in, year out, until it became accepted as fact. Remarkably, some of it lingers even today—the St. Swithin's saw is still aired regularly by the popular media in the twenty-first century.

To educated people, the idea that weather could be predicted was laughable. Writing on this subject in 1839, the *Gentleman's Magazine* bemoaned the fact that two popular almanacs, *Partridge's* and *Old Moore's*, "have professed, in the plainest terms, to foretell the weather, even to a day, stating that on one day there will be rain, on another snow, and on a third thunder."

The more worthy almanacs even satirized the weather and astrological predictions of their rivals. Early in the nineteenth century, for example, *Poor Robin* published the following "prophecy:"

> *Mars and Saturn are retrograde. This signifies that*
> *some strange country will be discovered, where the*
> *rivers run with Canary, the lakes and ponds filled*

with white wine and claret, the standing pools with muscadine, and the wells with pure hyppocras. The mountains and rocks are all sugar-candy, the hillocks and mole-hills loaf-sugar, fowls ready roasted fly about the streets, and cloaths [sic] ready-made grow upon trees.

Amid the flood of trashy almanacs there were a handful of credible publications that steered clear of making weather predictions, yet satisfied their readers' thirst for weather knowledge by engaging reputable men of science to write about weather in a knowledgeable way, based on what was actually known. One London publisher who bucked the trend of catering to the lowest common denominator was Charles Knight, who established the *British Almanac* in 1828. Another was George Cargill Leighton, who started publishing the *Illustrated London Almanack*, mentioned above. The latter employed the services of James Glaisher, who was superintendent of the Magnetical and Meteorological Department of the Royal Observatory at Greenwich. These more staid almanacs had less general appeal and lower circulation than their popular counterparts, but were appreciated by the better-educated middle classes.

By the 1830s, dozens of almanacs devoted to weather prediction had sprung up, and, though their predictions were about as reliable as—well, the weather—occasionally they would strike a lucky hit, and profits would soar. This was exactly what happened in 1838, when *Murphy's Weather Almanac* successfully predicted January 20 as the coldest day of the year. The day did indeed turn out to be a real bone-chiller, the coldest of the winter, with temperatures at Walton-on-Thames plummeting to -14°C. Patrick Murphy, its author, became a celebrity overnight, and sales of his almanac went through the roof. While critics in the scientific establishment sourly condemned his efforts as useless, and wags quipped about "prophets and profits," Murphy was laughing all the way to the bank.

All too often the weather prophecies in many of the almanacs were based on astrology, despite the fact that its practice was illegal in England, and their authors could not declare their astrological credentials for fear of prosecution.

Belief that the planets somehow exercise power over the weather dates back at least as far as Aristotle's *Meteorologica*, written in 340 BCE, and astrology remained a firm favourite of weather prophets for centuries. In 1686 a book called *Astrometeorologica* gave the world a new name for astrological foretelling of the weather, but even though astro-meteorology was already being dismissed at that time by men of science, its proponents wouldn't go away.

Of all the astrological weather prophets in nineteenth-century Britain, the most notorious was a man whose real name was Richard James Morrison, but who issued an almanac under the suitably prophetic-sounding pen name *Zadkiel Tao-Sze*, starting in 1836. His almanac was not as popular as *Old Moore's*, but still managed a healthy circulation of between twenty-two thousand and seventy thousand and thrived for almost seventy years.

In his early life Morrison had served as a lieutenant in the Royal Navy, and in February 1828, while posted with the Coastguard in Ardmore, County Waterford, Ireland, he led a rescue of a wrecked sloop. At great personal risk, he saved the lives of four men and a boy. He was subsequently awarded a medal for lifesaving from the National Institution for the Preservation of Life from Shipwreck for this courageous action, but it effectively ended his naval career. He had suffered from exposure during the rescue, which left him in poor health.

What drove a man like Morrison to astrology is anyone's guess, but, as Zadkiel, he was to become the foremost astrologer of his day, authoring five books on the subject. The fact that

Richard Morrison, a.k.a. Zadkiel Tao-Sze

he also had an interest in the occult should have raised enough red flags to prevent anyone's taking him seriously, but it didn't. Morrison strove to link astrological theory with meteorology, and he and the publishers of other similar almanacs were successful in implanting such a link in the minds of their readership; although one critic did condemn weather prophets like Zadkiel and his cronies as no better than fortune-tellers—adding sarcastically that they lacked the shrewdness, wit, and fun of that profession.

Morrison certainly went to a lot of trouble to gain acceptance in meteorological circles. He even managed in 1848 to worm his way into the presidency of the Meteorological Society of London, an on-again, off-again organization that struggled for years for survival, despite a list of eminent weather scientists among its membership. Morrison's presidency came to an abrupt halt in 1849, when he simply stopped attending meetings, but the very fact that an astrologer had become president may well have tarnished the society's reputation beyond any possibility of salvage. There is no record of the reasons for Morrison's sudden exit, but it's been suggested he was given the cold shoulder because his well-known association with astrology became anathema to a scientific society striving for respectability. The society collapsed soon after, re-emerged from the ashes, and (having shed its astrologer president altogether) finally succeeded as the British Meteorological Society in 1850. More than thirty years later it morphed again, into the Royal Meteorological Society.

But, unfortunately, the public perception of meteorology created by the almanacs—and by the astrologers in particular—made many truly scientific meteorologists distance themselves from the very idea that it might be possible to foretell the weather. And when real forecasts finally became a reality, they faced direct competition for the public's attention with the almanacs.

⌒⌒⌒

The almanacs were equally popular in North America, and publishers in both the United States and Canada developed their own home-bred versions. As in England, there were any number to choose from, and they varied enormously in quality—and in the extent to which they retained a foothold in reality.

Benjamin Franklin himself published *Poor Richard's Almanack* for twenty-five years. He did give weather predictions, although he steered clear of astrology, and some say that he did not expect anyone to take them seriously. Robert B. Thomas's *Farmer's Almanac*, published in Boston, catered to an audience that was generally educated and intelligent, and avoided astrology and superstition, but did include advice on conducting agricultural affairs at certain phases of the moon, perhaps with tongue in cheek.

One of the most respectable Canadian titles was the *Canadian*

Almanac, and Repository of Useful Knowledge. First published by the firm of Scobie & Balfour in 1847 in Toronto, it provided real astronomical information "made expressly for this publication at the magnetic observatory in Toronto." It gives the moon's phases, and a monthly meteorological table with mean high and low temperatures to be expected for each month for Upper and Lower Canada, and in 1848 stated its willingness "to extend these to Nova Scotia, New Brunswick and Newfoundland if reliable information is available."

Other Canadian offerings were less concerned with hard facts, at least in their weather information. One such, *Bryson's Canadian Farmer's Almanac*, was published in Montreal. The almanac gives month by month predictions. In December 1851, for example, the Canadian farmer should expect:

> *Cold days continue to the next full Moon on the 8th, followed by Cold with high winds. But a small depth of snow until the 15th. After which there will be a tremendous snow storm and cold weather will last the remainder of the month.*

Well, yes…and after that expect a visit from Santa Claus.

Yet other almanacs started out with good intentions. *Belcher's Farmer's Almanac*, for example, the product of Clement Horton Belcher, a prosperous Halifax bookseller and publisher, first appeared under that title in 1832, and survived for almost a century, becoming a feature in most literate Nova Scotian households. Belcher focused primarily on giving sound advice on modern farming, but eventually public demand drove him to include weather predictions.

The publisher of another almanac, Halifax printer and newspaperman William Cunnabell, clearly put little store by the weather predictions he included, and tried to distance himself from them.

BELCHER'S

Farmer's Almanack

FOR THE YEAR OF OUR LORD

1869:

PROVINCE OF NOVA SCOTIA,

DOMINION OF CANADA,

Being the First after Bissextile or Leap Year, and the latter part of the THIRTY-SECOND and beginning of the THIRTY-THIRD Year of the Reign of

HER MAJESTY QUEEN VICTORIA.

[ESTABLISHED IN 1824.]

PUBLISHED BY THE PROPRIETOR

C. H. BELCHER,

HALIFAX, NOVA SCOTIA,

SOLD BY

A. & W. MACKINLAY, No. 10 GRANVILLE STREET,
R. T. MUIR,.......... " 125 " "
M. A. BUCKLEY,..... " 85 " "
Z. S. HALL,.......... " HOLLIS STREET.
CONNOLLY & KELLY, " 36 GEORGE STREET

1869.

INDEX WILL BE FOUND ON PAGES 191 AND 194.

Cunnabell's Nova Scotia Almanac and Farmer's Manual for 1842 includes a "Table of the Weather For Every Day of the Year," followed by a disclaimer:

> *A correspondent has furnished the preceding table, which he assures us is founded on an approved theory. We thank him for the pains he has thus taken on our behalf, and hope his expectations may be realized with regard to the discovery; but without questioning his intelligence, we would here intimate, that should his calculations turn out contrary to what he expects, he may console himself with the reflection, that when such men as Sir W. Herschel, and Dr. Adam Clark were disappointed in the results of a similar table, his has not been the only failure.*

These publishers were, after all, businessmen. If they were to sell their products, they had to give the public what it demanded.

***

In the same way that weather has forever occupied the minds of humankind, the belief that the moon holds a steady influence over its affairs, and those of nature, has been around since antiquity, and no doubt beyond. Indeed, there is probably enough folklore surrounding the moon to fill a book. Yet, while the moon's influence over the tides is well and truly established, there has never been the merest jot of evidence to support any of its other proposed influences. And there have been dozens.

According to eighteenth- and nineteenth-century moon seers, it is advantageous, for example, to sow hemp and flax at certain phases of the moon, to sow or plant certain vegetables (but not others) and fruit trees as the moon is waxing, and while it waxes also to cut firewood (to prevent snapping when the wood burns), to sow wheat without fear of its smutting as it ripens, and to destroy pernicious weeds. Apples picked on a full moon were said to keep well throughout the winter.

Others believed that predictions can be made from the moon's appearance, or from whether it casts a shadow; that moonlight darkens the complexion; that crabs, oysters, and other shellfish are larger during the waxing than the waning moon; and that moonlight hastens the decomposition of animal substances. As late as 1904, author George Kittredge wrote that there were mothers in New England who still cut their children's hair when the moon was waxing, in the belief that it would grow back more luxuriantly, and that some farmers were still following an ancient rule of killing their pigs as the moon was waxing so that the meat would swell, rather than shrink, during pickling or cooking.

Even at the end of the eighteenth century physicians were maintaining that an approaching full or new moon was "a powerful exciting cause of fever," and that very old folks usually died at full or new moon. Diseases that were influenced by the moon included ulcers, measles, spots on the face, cataracts, epilepsy, and dysentery. Astrology also held that every organ or part of the body was governed by a sign of the zodiac, and the physician needed to know whether the moon was in or out of that sign before treating an ailment affecting that part. And of course every school child learned early that a full moon brings on madness, and may even turn humans into werewolves!

It was widely believed that weather events could be attributed to the phases of the moon, and this was a perpetual favourite with the almanacs. That myth had been around since at least the second century AD, when the astronomer Ptolemy made a number of remarks on how the moon's course and appearance provide indications of future weather—"If she appear dark, or pale and thick, she threatens storms and showers," for example.

Some almanacs produced weather tables based on the hour of the day or night when the moon changes its phase. *Bryson's Canadian Farmer's Almanac* for 1851 includes, for example, allegedly at the request of its readers, a weather table "constructed from considerations of the attraction of the Sun and Moon, in their several positions respecting the earth, and by experience of many years' actual observations," giving weather predictions at various phases of the moon. According to this table, if the new and full moon occur between the hours of 12 noon and 2 p.m., the reader should expect very rainy conditions in summer or snow and rain in winter.

The British *Farmer's Almanac* went so far as to falsely claim the well-known scientist Sir William Herschel as the author of its predictions, which were total fantasy, but being associated with such a distinguished name, the predictions remained popular among farmers for many years. And then there was the notion, championed by the otherwise respectable meteorologist Luke Howard—the man who gave us the names for different types of cloud, such as cumulus, stratus, and cirrus that form the basis of the way we classify clouds today—who incorrectly insisted that the moon's declination (that is, how far it was positioned north or south of the celestial equator—how high in the sky it appears) had an effect on average rainfall and barometric pressure.

Amidst all the lunar nonsense there were, however, a few glimmers of truth. There is some substance, for example, in the belief that the appearance of the moon—coronas, halos, lunar rainbows, or a watery appearance—can indicate that wet weather is imminent. The reasoning—that these indicators are caused by water vapour rising into the upper atmosphere—is sound. It is like so many other weather portents that have become the stock-in-trade of farmers and seafarers, based on centuries of observation. There's a clear distinction between weather portents developed from experience (*"Red sky at night, sailor's delight. Red sky in the morning, sailor's warning,"* for example) and the weather saws propounded by the almanac publishers. The weather *portents* are usually based on tried and tested observations born out of centuries of practical observation by those who depend for their livelihoods or even their safety on an understanding of weather patterns and their vicissitudes. The weather *saws* (such as the one about St. Swithin) generally have no substance at all.

There was one attractive theory concerning the moon and weather, however, that seemed, on the face of it, perfectly reasonable.

Although there is no truth in it, the "Lunar Theory" of weather does not sound altogether unlikely. In fact it sounds almost intuitive, for the theory's underlying argument is that if the moon is mainly responsible for the rhythmic rise and fall of the tides in the oceans, as has been known since antiquity, then why should it not cause disturbances or tides in the atmosphere, too? There are a number of similarities between the atmosphere and the oceans. Both are fluids—the atmosphere a mixture of gases and the oceans liquid; both surround the earth (the oceans more or less); and both are in constant motion. And if the moon does cause disturbances in the atmosphere, then why might it not have a similar effect on the weather?

In fact the moon does exert a rhythm on the atmosphere, similar to that of the oceans' tides, but scientists had recognized long ago that the effect was so miniscule as to be insignificant. Although it had very few supporters in the scientific community, the lunar theory did have some followers. They were actually hoping to establish real scientific credibility for their theory, but it had become associated—albeit incorrectly—with astrology. Astrologers, regarded in well-educated circles as being on the lunatic fringe, were eager to grasp anything associated with the movement of heavenly bodies. They also claimed that, in addition to the motion of the moon, the atmosphere was influenced by the motion of the other planets and the sun. Needless to say, the principles applied by the astrologers had no scientific plausibility whatsoever.

Nevertheless the lunar theory itself was beguilingly attractive. It was an idea that certainly beguiled Stephen Saxby, and one that was soon to become linked to his name.

~~~

It was six decades after Franklin's great discovery about the progressive movement of storms before there was another significant breakthrough in our understanding of these events, and it too took place in America. Nowhere else in the Western world are there storms with such immense power as hurricanes, and it is in the Caribbean, the Gulf of Mexico, and along the eastern seaboard of the United States that they most often unleash the worst of their fury. It's true that storms with similar force are felt in Europe, but they seldom reach the intensity of a major hurricane. It's hardly surprising that hurricanes arouse particular interest in North America.

The breakthrough was made by a man who seems an unlikely candidate for scientific greatness. Born of English parents, in rural Middletown (now Cromwell), Connecticut, in 1789, William Redfield decided to add an extra initial to his name and become William C. Redfield. When asked what the "C" stood for, he would answer "Convenience," which suggests he had a dry sense of humour, was a bit of an oddball, or possibly both. In fact, he inserted the "C," or so he said, purely to distinguish himself from two other William Redfields in his hometown. It was indeed for convenience.

When William C. was thirteen years old his seafaring father died, leaving the family in straightened circumstances. Shortly afterwards the boy was apprenticed to a saddle and harness maker in his hometown. During his little spare time, William studied hard with the help of a local physician who had befriended him, realizing the boy was extremely intelligent, and had lent him books from his extensive library. Two years later William's mother remarried and moved to Ohio, along with his five younger brothers and sisters.

In 1810, when he had completed his apprenticeship, William set out—on foot—to visit his mother, an epic journey of some seven hundred miles. It was the kind of trek that no one today would even contemplate without months of preparation, a support team, sponsorship, and a camera crew to record the whole adventure, but one that Redfield seemed to regard as little more demanding than a Sunday afternoon ramble.

Redfield and two companions laced up their boots and made the trip in twenty-seven days, trudging for much of the trip through dense forests—there were few roads, and where roads did exist the trio couldn't afford to ride the stagecoach. On the southern side of Lake Erie, the young men found that the only easy way to travel was along the lakeshore. After his visit, when it was time to return home, Redfield donned his boots again, and without a second thought trudged all the way back to Middletown.

It was during this trek that Redfield had his first encounter with steam navigation and, like the young Stephen Saxby in England, the American Redfield was inspired by the sight. The vessel he saw in Albany in New York State was one of the earliest steam-powered ships ever to operate commercially in North America. It was Robert Fulton's boat, the *Clermont*. The trek also offered wonderful possibilities for observing nature, and it was as an observer that the largely self-educated Redfield excelled, jotting down his observations in a journal.

On his return Redfield plied his trade for ten years as a mechanic, also keeping a small store, and then in 1822 he went into business as a steamboat operator on the Connecticut River.

The boilers in the early American steamboats, like those in Britain, exhibited an alarming tendency to explode. Such explosions were so frequent that the travelling public became leery of steam locomotion, but the businessman Redfield found a simple way of alleviating their fears. He introduced a tug and barge system. The passengers would travel in the comfort and security of a "safety barge," at some distance from the hazardous boiler. The crew—presumably expendable—were the ones who risked life and limb as they towed their passengers hither and thither along the Hudson River between New York and Albany. The venture did not last long. Eventually the travelling public lost its fear of steam boilers and self-contained passenger boats became the norm once again, so Redfield, a canny businessmen, converted his barges to carry freight and operated them successfully for many more years.

Eleven years after his epic journey on foot, Redfield made another journey, during which he made the discovery that would ensure him a place in the history of weather science. In 1821, a furious hurricane, originally known as the Great September Gale, and later as the Great Norfolk and Long Island Hurricane ,devastated much of the eastern seaboard from the Carolinas to Massachusetts, damaging buildings, tearing down trees, and causing widespread flooding. In New York City it caused a large and rapid storm surge that flooded the city's wharves. Fortunately, it occurred at low tide, or the flooding would have been far worse. Shortly after the hurricane had ripped through Middletown (and New York City) on September 3, Redfield happened to be on the road again. Near Middletown he noticed that trees felled by the storm were all blown down toward the northwest. But when he arrived in western Massachusetts, about a hundred miles to the northwest, he noted that trees there had been blown down with their tops pointing in exactly the opposite direction, toward the southeast.

William C. Redfield, the man who discovered that winds in a hurricane rotate around its centre.

Redfield thought about his observations for several months before coming to the inescapable conclusion that if the winds had been blowing in opposite direction in the two places at the same time, then the storm must have been like an enormous whirlwind.

He was not the first person to notice this feature of hurricane winds. Credit for the original idea is usually given to William Dampier, the English buccaneer turned explorer. In 1687, he had described a typhoon in the China Sea (the eastern equivalent of the western hurricane) as a "sort of violent whirlwind," but he had not developed the idea.

At the time, Redfield did not consider himself anything more than an interested amateur. He gave no thought to publishing his

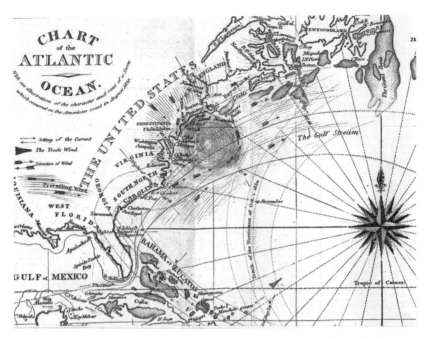

Part of a map by William C. Redfield showing the progress of an 1830 hurricane.

discovery, and so he kept it to himself. It would have gone altogether unnoticed (although sooner or later someone else would inevitably have stumbled to the same conclusion) if not for a chance meeting with a man called Denison Olmsted, a professor at Yale College, who was travelling on one of Redfield's steamboats from New York to New Haven. During the course of their conversation Redfield mentioned his observation, and Olmsted immediately realized its significance. He persuaded Redfield to publish his insight in the scientific literature, in the *American Journal of Science*. It was the start of Redfield's scientific career.

Redfield pursued his interest in storms—North Atlantic hurricanes in particular—with renewed vigour, even though as a businessman he could devote only his spare time to their study. He now turned to another source that would help him confirm the picture that he was piecing together. He collected a great number of records of storms at sea, as made on the spot by a number of sea captains who sailed out of New York and sent him extracts from their log books. He culled more data from marine reports of vessels that arrived in port after storms, and he corresponded with ships' masters to obtain their written statements.

In this way he investigated hundreds of storms. For one hurricane alone—the Cuban hurricane of October 1844—he gathered information from no less than 164 sources. From all the data he laboriously analyzed, he was able to plot on marine charts the storms' dimensions, their speed of rotation and speed of progressive movement, and the wind strengths at different distances from the storm centre.

From his observations, Redfield noted that the winds always circulated in a counter-clockwise direction around a storm centre in the northern hemisphere, and he correctly predicted that they would rotate clockwise in the southern hemisphere. He published many more articles on storms, mostly in the *American Journal.* Everything he observed seemed to confirm his original conclusion that storms consist of a rotating system of winds, and that (as Franklin had already shown), they move progressively from one place to another. A few years later, Redfield was also able to show that hurricanes were both long-lived and travelled enormous distances.

Perhaps because, as an engineer, he had a very practical nature, Redfield was not content to simply add to scientific knowledge. He wanted to put his discoveries to practical use. Redfield's work caught the attention of William Reid, a British officer with the Royal Engineers, who had encountered a deadly hurricane in Barbados in 1831 that had killed almost fifteen hundred people, and he too had gathered as much information about this storm as he could from ships' logs. Reid's work supported Redfield's, and for many years the two men shared information about tropical storms. From their studies they were able to compile guidelines or rules for navigators, which told how to steer clear of the most dangerous parts of a hurricane. These guidelines were published in the *American Coast Pilot*, and undoubtedly saved many lives at sea.

At the same time, another British sailor, Henry Piddington, a captain with the East India Company who was studying typhoons in the Pacific and Indian Oceans, was inspired by the work of Redfield and Reid. He coined the name "cyclones" for these storms, and his main concern also was improving safety at sea. Piddington took up his pen and wrote a practical book, *The Sailor's Horn-Book For the Law of Storms*. The "horn" in the title referred to a thin slice of translucent horn, which came with the book and was engraved with a diagram of a cyclone and the points of the compass. When placed on a chart, it gave the mariner an indication of the probable wind direction in an approaching storm, from which he could figure out what course to steer to escape the worst of its violence. The book became a classic, the standard work for mariners on storms.

There is a simple explanation for the guidelines that Redfield, Reid, and Piddington concocted. The winds in an Atlantic hurricane or a Pacific cyclone in the northern hemisphere blow around the storm's centre in an anticlockwise direction, at the same time as the hurricane is moving forward, nearly always in a northerly direction. Now, if the winds to the right of the storm's centre are blowing from the south at, say, eighty miles an hour and the storm is moving north at twenty miles an hour, the effective wind speed is one hundred miles an hour. Meanwhile, to the left of the storm's centre, the winds are blowing from the north, also at eighty miles an hour, but because of the storm's northward passage at twenty miles an hour the effective wind speed in this part of the storm is only sixty miles an hour. That's a difference of forty miles an hour, and it is a huge difference, because of the relationship between a wind's speed and its force. As the wind's speed doubles, its force and therefore its violence is not simply doubled, but quadrupled. Piddington recognized that, in the northern hemisphere, the storm's right-hand side is the one to avoid. The "dangerous semicircle" he called it, the left-hand side being the "safe semicircle."

The word "safe" here is of course purely relative; there is no safe place to hide in a hurricane in the open ocean—just those where your chances of survival are slightly better.

✺

Redfield had shown *how* the winds in a hurricane blow, but he was neither particularly interested in, nor did he make any serious attempt to figure out *why* they blew that way, or what forces of nature were at work. He believed rather simplistically that hurricanes were born where the easterly trade winds meet the islands of the Caribbean and form whirls, as when water in a shallow stream meets a rock and forms a whirlpool. He rightly discounted one popular theory—that electricity in the atmosphere was the cause of storms—but he wrongly believed that heat played no part, and (when pressed) offered instead his own somewhat half-baked explanation. As will become clear, he would have saved himself a lot of grief if he had stuck to describing storms, rather than theorizing about what causes them.

The full picture would not emerge until well into the twentieth century, but by the early 1830s another man of science, James Pollard Espy, was already beginning to consider the possibility that somehow heat was involved in generating the terrific forces that are unleashed in storms. Unfortunately, his ideas were about to unleash a storm of their own in the United States, a storm of controversy that would rock the burgeoning science of meteorology to its foundations.

CHAPTER 2

A Storm of Ideas

Meteorology has ever been an apple of contention, as if the violent commotions of the atmosphere induced a sympathetic effect in the minds of those who have attempted to study them.

—Joseph Henry, 1858

When Stephen Saxby—who as a boy had been fascinated by the sight of the steamboat *Thames*—was in his mid-thirties, he brought his wife and four children to the picturesque village of Bonchurch on the Isle of Wight, a short ferry ride across the Solent from England's south coast. By now a typical middle-class Victorian gentleman, Saxby, with his many interests, was among other things an avid fossil hunter, and on the island he was in his element. There is no place richer in fossils in the British Isles, and it remains a favourite haunt of fossil hunters to this day. The Isle of Wight is best known for the many dinosaur bones that have been found there, dating from the Cretaceous period, but it also contains a host of fossils from other animal phyla—reptiles, fish, and invertebrates—spanning that period of almost a hundred million years.

The story of Stephen Saxby's early life and career have been lost with the passage of time. We do know that, as a young man, while still living in Ramsgate, he was part of a lifeboat crew and took part in a number of rescues from ships wrecked on the nearby Goodwin Sands, a notorious deathtrap for shipping in the English Channel. His encounter with the *Thames* is the only other event of which we can be certain until 1827 when, aged twenty-three, he married Mary Ann Lindeman, and for the next twelve years or so the couple seldom lived in one place for very long. Four years after they were wed they were in Middlesex and had started a family. They had two sons and a daughter in that county and then moved to Eltham in Kent, where another son was born. There are suggestions that Saxby spent some of his younger years as a mariner, and that for some of those years he was working for the East India Company, but by the time the family moved to Bonchurch he had embarked on a career in teaching. It was a vocation he would pursue for the rest of his working life.

Saxby came to Bonchurch to teach at a school in a local villa named Mountfield. Far from the grime and the smoke of London, and situated between Shanklin and Ventnor on the island's eastern shore, Bonchurch has spectacular views of the English Channel, and for a well-to-do teacher and his growing family in the middle of the nineteenth century it was a comfortable setting. The couple had two more children soon after; another son and another daughter.

Bonchurch became something of a haven for writers and artists in the nineteenth century. It is the place, for instance, where Charles Dickens rented a country house called Winterbourne while he wrote his great novel *David Copperfield*, sometime around 1848. Its climate is pleasing and, according to contemporary island historian William Davenport Adams, is "beneficial in pulmonary diseases."

Much of the Island's coastline consists of towering chalk cliffs,

The village of Bonchurch, Isle of Wight, in the mid-nineteenth century.

rich in fossils. The constant erosion from wind, salt spray, and sea is always exposing new specimens to the eye of the fossil hunter. Beneath the chalk there are layers of greensand, a kind of rock with very little strength that gets its green colour from an iron-bearing mineral. About ten thousand years ago, as the ice that had covered Europe for centuries retreated, large areas of these greensand cliffs broke loose and slid into the sea. These enormous landslips stretch for several miles, from Blackgang in the west to Bonchurch in the east, forming what is known locally as the Undercliff, and this lucky accident of geology reveals even more fossils that once lived at the bottom of an ancient sea.

Saxby spent much time of his spare time scouring the shore and Undercliff for fossils, and he was able to put together a valuable collection. He found his best specimens in the greensands, but, being an amateur, he was not able to identify them and sent them to an expert for description and identification, the eminent geologist Dr. William H. Fitton. During his rambles along the island's shores he discovered two fossil species that were new to science. One, a close relative of the modern lobster, was named *Hoploparia saxbyi*, after its discoverer; the other, a relative of octopus and squid, was a chambered nautilus that was named *Nautilus saxbii*.

It seems that Saxby was also dabbling in archaeology, as in 1847 he discovered, in the local church of St. Boniface, several murals that had been covered in whitewash for centuries. That same year he won an archery contest at the island's Carisbrooke Castle, using arrows he had designed himself, astutely allowing for their centre of gravity.

This was one of the most settled periods of Saxby's life, as the family remained on the island for about ten years. It was also an extremely busy one. In addition to his teaching, Saxby was engaged by the marine insurance company Lloyds of London as an agent—a person living in a port who managed the business of the company's members and sent in reports of shipping movements and accidents. In this position he was well aware of the appalling number of ships that were wrecked around the coast of Britain, mostly as a result of violent storms, of whose approach mariners—and hereabouts fishermen in particular— could expect almost no warning whatsoever. If he had not begun to do so before, we can be sure that by now he was turning his attention to ways of improving safety at sea.

∾∾∾

If anyone ever compiles a list of the most preposterous scientific proposals of the nineteenth-century—and it would be a long list—it's sure to include the suggestion that the western forests of the United States should be put to the torch on a massive scale—not just once, but repeatedly—so that during a drought the eastern part of the country might benefit from some rainfall. Fortunately for Americans living out west, the scheme—dubbed "crack-brained" by no less than former United States President John Quincy Adams—was never put to the test.

Oddly enough, this outlandish idea emanated not from the brain of some cabin-crazed backwoodsman, but from the mind of a brilliant scientist, James Pollard Espy, and it is probably as well for his fellow countrymen that he devoted most of his time to thinking about storms, rather than to tinkering in the woods with a box of matches and a pile of tinder.

For, rainmaking ventures aside, it was Espy who produced the next major insight into the nature of storms—an odd achievement for a man who started out with no real interest in science. After completing his education at Transylvania University in Lexington, Kentucky, he taught mathematics and classics in Ohio, Maryland, and Philadelphia, and also studied law. Then he converted to science, joined the Franklin Institute in Philadelphia, and began to study the atmosphere. He published his first paper in 1821, then his interest broadened to include meteorology, and storms in particular. Eventually, and despite his preposterous notions on rainmaking, Espy played a key role in the effort to establish a weather forecasting system in the United States.

Unlike Redfield, Espy made his great discovery not in the great outdoors where weather and its effects can actually be observed, but in his laboratory, where he built a device he called a "Nephelescope," a chamber in which he could replicate the behaviour of the vapour-laden air within clouds. From his laboratory experiments he deduced in 1833 that at the centre of a storm the air rises, expanding and cooling as it does so until its water vapour condenses to water and starts to fall as rain (at the temperature known as the dew point). This process of convection, he believed, was the driving force that gives storms their enormous energy. It's sometimes described as the "chimney effect."

Espy revealed another secret of clouds: when the water vapour in a cloud condenses it releases heat, which makes the surrounding

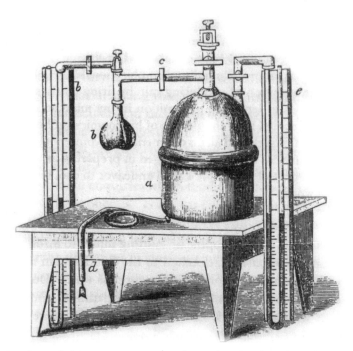

James Espy's Nephelescope, a contraption he used to simulate cloud formation in the laboratory.

air expand and rise even higher. Physicists refer to the heat in the rising air as "latent heat," meaning "hidden" heat. It is hidden because when water changes from liquid to vapour form—by evaporation from the ocean's surface for example—it takes up a lot of heat, but the water molecules themselves do not increase in temperature. This is the heat, locked up in the water vapour, that is later released when the vapour condenses. The warmth drives the system still higher—and it is this upward movement of air and clouds that give a storm its energy. As the air rises, air pressure at the earth's surface is lowered, and this in turn draws in air from the surrounding area in the form of wind. That's why a hurricane, once it starts to build, becomes self-sustaining. It is a heat engine that will continue to generate enormous power in the form of wind as long as it is supplied with a fuel source—warm, moist air. When it reaches cool northern latitudes its fuel supply is generally reduced, and eventually cut off altogether, so that the storm eventually withers and dies.

But although Espy had discovered an important underlying principle of storm formation, his notion of the direction that winds actually blow in a storm was quite wrong. Winds, in Espy's opinion, blow directly *into* the centre of a storm or any other area of low pressure. This, of course, completely contradicted Redfield, whose observations of the same events showed that winds blew *around* a storm centre. And when Espy came across Redfield's articles on storm systems in 1834, he set in motion a dispute that polarized the ranks of meteorologists.

Espy was, by most accounts (even according to his friends), a difficult man. Although a brilliant scientist, he possessed a stubborn streak. Once he had settled on a theory, his faith in his own opinions was unshakable, even in the face of evidence that did not support his ideas. He was extremely argumentative, highly critical—even contemptuous—of the work of colleagues, a habit that didn't win him many friends in scientific circles, yet he disliked any criticism of his own work. He did have at least one endearing feature, however; he shared Benjamin Franklin's fondness for flying kites, and you can forgive a man a lot for that.

Redfield and Espy approached the problem of understanding storms from entirely different viewpoints. Whereas Redfield made careful observations and then developed a picture of the way they worked, Espy came up with a theory and then tried to make the facts fit his theory. The two approaches are both valid scientific methods, although the proponents of each seldom like to admit it.

Who would have thought that the weather could draw grown men into such bitter disagreement that they would end up going at each other, hammer and tongs, for the best part of a decade? The long-drawn-out dispute involved a number of issues, but in essence it boiled down to two major points of disagreement. Redfield would not accept Espy's theory that the heat of the sun played a major role in the creation of storms, and Espy would not accept Redfield's interpretation of the pattern of their wind flow.

It was Espy who fired the first salvo in what eventually became known as the "American Storm Controversy." Although he admitted that Redfield's articles contained some valuable ideas, he added that "some of them, however, are so anomalous and inconsistent with received theories that I hesitate to put entire confidence in them, and shall continue to doubt until I have the most certain evidence of the facts."

Redfield was not going to take this lying down. He was incensed by Espy's dismissal of his description of winds rotating around a storm centre because they did not conform to current "received" theories. He *knew* he was right because he had made thousands of meticulous observations that supported his ideas. He responded with indignant surprise at Espy's denial of the evidence: "I did not anticipate so complete an evasion of all the distinguishing points at issue, and so barren an effort at confusing and mystifying the most distinct phenomena of this storm."

To be fair to Espy, he objected to Redfield's theory because, now that he had hit on a key principle underlying the driving force of storms, Espy could imagine nothing that would cause winds to blow around a storm, rather than to its centre. The idea was counter-intuitive, and, to Espy, Redfield's observations seemed to defy logic.

And now, unfortunately, Redfield undermined his own position by bowing to pressure to develop his own theory on the nature of storms. Scientific theories and laws were the stuff that enhanced scientific reputations at this time, and Redfield was persuaded by the astronomer and mathematician Sir John Herschel, in England, who had become an ally, that a grand theory would make his colleagues sit up and take notice. And so Redfield came up with a vague assertion that gravity or "the dynamics of the atmosphere" were the driving forces of storms. It was far from convincing, and left him wide open to further criticism.

Initially both sides behaved civilly, on the surface at least, and exchanged courtesies and compliments before politely criticizing each other's viewpoints, but soon the disagreements degenerated into a verbal brawl. Most of it took place within the pages of two learned journals—the *Journal of the Franklin Institute* and the *American Journal of Science* (in which Redfield had published his first article). There was also much correspondence between each of the opponents and their supporters. Accusations of misquoting each other, of trying to confuse the issue, and of appropriating and manipulating data flew back and forth.

In 1837 Espy, a great self-promoter, began a series of speaking engagements on the American Lyceum Movement lecture circuit, which gave him a platform from which he could take further jabs at Redfield's ideas, and he took every opportunity to do so.

Redfield himself never went on the offensive in public, preferring to confine the debate to the pages of the journals, and he never engaged in a public showdown with Espy. A quiet, diffident man, it would have been out of character for him to do so. It was left to the man who had first brought him into the scientific fold, Denison Olmsted, to take up the public fight, much in the same way that Thomas Huxley battled those who poured scorn on Charles Darwin after the publication of *On the Origin of Species.* Olmsted went head-to-head with Espy on at least two occasions, in New York and Boston.

Redfield and Espy were by no means the only protagonists in this dispute over the character of storms and the nature of the forces involved in their creation. Each had his own supporters, and the battle lines were drawn to some extent by geography. Redfield's bastion of support was based in New England; Espy's in Philadelphia.

For the public it was a marvellous spectator sport and it made excellent copy for the gentlemen of the press, who reported on every verbal brickbat with relish. They found great material in Espy, who used his lecture tours to promote his bizarre rainmaking ideas, earning himself the titles "Storm King" and "Storm Breeder." The publicity increased his standing in the public eye, though not in the eyes of his fellow scientists.

A third combatant now decided to enter the fray. Robert Hare, a chemistry professor at the University of Pennsylvania, announced his own pet theory that electricity was involved in the generation of storms, and he attacked Redfield because the latter ignored this possibility completely. Hare initially sided firmly with the Espy camp, but that soon changed when Espy, with his usual talent for diplomacy, called Hare's theory unphilosophical. The idea that electricity in the atmosphere was somehow responsible for storms did have its followers, but no one was convinced by Hare, and he was soon pushed to the sidelines.

The dispute became as well known in Europe as it was in the United States after it received an airing, in 1840, at the annual meeting of the British Association for the Advancement of Science, held in Glasgow that year. Redfield and Espy were both invited. Redfield was unable to attend, but sent copies of a couple of papers, which arrived too late to be read at the meeting but were later published in England. Espy did attend, but his presentation was received coolly by the British scientists, perhaps because Redfield's ideas already had the support of members of the British scientific elite, among them Redfield's friend Sir John Herschel. Undeterred, Espy travelled to France, where he received a warmer reception by the scientific establishment and made allies of the great astronomer and physicist François Arago and others.

The dispute had cooled to a simmer by the time Espy was appointed official meteorologist (not to be confused with the modern understanding of the term—forecasting still lay almost two decades in the future) to the War Department in Washington in 1842. He discontinued his public lectures and did not give another until 1849, although he did use his new position of influence to further pour vitriol on Redfield's theory. Redfield resented Espy's appointment, with some justification, seeing it as official acceptance of the latter's ideas, at the expense of his own. Nevertheless, both men became focused on their work, rather than on the dispute, for the next few years.

If only Espy and Redfeld had restricted themselves to what they did best—Redfield the brilliant observer, and Espy the great theoretician—the whole ugly situation might have been avoided, but they didn't. Each had grasped one valid piece of the puzzle, but failed to realize that his opponent had grasped one too. But there was to be a positive outcome of the sorry affair. Meteorology was now front and centre on the public stage, and the dispute had motivated a number of other men of science to try to resolve the dilemma of storms.

Redfield was right. Winds do rotate around a storm's centre—but not quite in the way that he believed. Unknown to both Redfield and Espy, the explanation lay in a paper published in 1835 by a French professor of Mechanics at the École Centrale des Arts et Manufactures (now the École Centrale Paris) called Gaspard-Gustave de Coriolis, a name that lives on in the curious phenomenon he discovered known as the Coriolis effect.

If the earth were stationary, rather than rotating on its own axis, then winds would blow directly toward the centre of a storm, as Espy insisted they do—blowing almost directly from areas of high pressure toward those of low pressure, as when you let a balloon deflate. But the earth is not stationary, of course, and its rotation causes the winds to deflect. Imagine you are standing somewhere in the northern hemisphere and are able to toss a ball, not just a hundred metres or so, but a few hundred kilometres. It would not travel in a straight line, but would be deflected to the right.

The same thing happens to the air molecules of the atmosphere. When winds are blowing across the rotating earth's surface they too are deflected to the right in the northern hemisphere and to the left in the southern. So when wind starts trying to get from an area of high pressure to one of low pressure, in the northern hemisphere it is deflected to the right, and ends up blowing around the low pressure centre (and slightly inwards).

Coriolis was not, it turns out, the first to observe this effect. Almost two hundred years earlier, in 1651, an Italian scientist called Giovanni Battista Riccioli published his opinion that if the earth were in motion, a cannon ball fired toward the north would be deflected to the right in the northern hemisphere, but Riccioli and his assistant, Francesco Maria Grimaldi, got things backwards—they thought that this did not actually happen in practice (although it would be observed later). They incorrectly took this supposed lack of deflection as evidence that the earth is stationary, in support of their belief that the earth is at the centre of the universe and is orbited by the sun, moon, and stars.

And although the effect bears Coriolis's name, he was an engineer, concerned primarily with mechanical issues—the rotation of water wheels and the like. He had no interest in pursuing the relevance of his discovery to the motion of the atmosphere, although he did hint that it might be important. It was another two decades before anyone would follow up on that hint. And when you consider that Coriolis described this effect—in the arcane style much beloved by scientists, but which leaves the rest of us glassy eyed—as "motion in a rotating frame of reference," it's hardly surprising.

Matthew Fontaine Maury

On October 17, 1839, a stagecoach accident forever altered the course of meteorology. An ambitious thirty-three-year-old U.S. Navy lieutenant called Matthew Fontaine Maury was riding the stage from Fredricksburg, Virginia, to Tennessee to visit his parents, whom he had not seen for several years. In the middle of the night, just outside Lancaster, Ohio, the overloaded and top-heavy coach took a curve too quickly, went off the road, and flipped over. Maury, who was travelling on top of the coach, was thrown clear of the wreckage and hit the ground hard. His thigh-bone was fractured in two places and his knee dislocated.

It was a crippling injury—the first attempt to repair the bone failed and it had to be painfully broken again (without anaesthetic) and reset. Even after the leg was patched up, the lieutenant's convalescence was slow—so slow that he was unable to report, as ordered, for duty in New York aboard his new ship, the brig *Consort*. The injury had disabled his leg so badly that he was no longer fit for active service at sea of any kind.

It was touch-and-go for Maury for thirty months before the navy found him a suitable niche. Before his accident he had written a couple of articles and a textbook on navigation, and his scientific ability had led to his selection as astronomer to the U.S. Exploring Expedition of 1838 to the Pacific region (often called simply "the Ex-Ex"). Maury declined, which was probably a wise decision, as the expedition was eventually led—disastrously—by an incompetent martinet called Captain Charles Wilkes.

Maury's fortunes improved in 1842, when he was appointed to head the U.S. Navy's Depot of Charts and Instruments in Washington (a position once occupied by Wilkes), where the navy maintained and adjusted chronometers, using astronomical observations, and stored other instruments and its supply of charts for its ships. Two years later the depot became the location for a new U.S. National Observatory.

Despite Maury's interest in scientific matters, he had little aptitude for or interest in astronomy, but in his new appointment he made a discovery that really did interest him: a hoard of ships' logbooks that had been gathering dust in the old depot. Maury realized he was sitting on a treasure trove of data that could be put to good use, and he embarked on a study of the winds and currents of the world's oceans. By painstakingly extracting data from the logbooks and collating it, he was able to start compiling a collection of charts that showed the patterns of oceanic winds and currents, starting with the Atlantic Ocean. He was employing the same laborious technique that Redfield had earlier used to chart the track of hurricanes, albeit with a staff drawn from the ranks of the navy, mainly midshipmen, to help with the mind-numbing tedium of the task.

When Maury published his *Wind and Current Charts* in 1847, they were an instant success with mariners. First, Maury made them available for free—in exchange for further information on winds and currents from the captains' logbooks. And second, by using the charts to find the most favourable winds and currents, sailors could often shave much time from long ocean passages.

They also established Maury's reputation as an expert on the weather.

Cambridge University has acquired many strange traditions during its long history. Among them is a curiosity called the "ten-year man," a dispensation dating back to the reign of Elizabeth I that allowed a man over the age of twenty-four to register ("matriculate") with the university and, as long as he participated in certain academic exercises that tested his powers of reasoning, after ten years to proceed to the degree of Bachelor of Divinity without taking any previous BA or MA degree. He was not required to reside at the university except for three terms during the final two years. The scheme was little used until the mid-nineteenth century, but then there was an upsurge in its popularity.

Stephen Saxby was one man who took advantage of this option. He had left part of his formal education—the gaining of academic credentials—until rather late in life, but in June 1845, he entered

Gonville and Caius College, Cambridge (often called just "Caius College," and pronounced "Keys"), as a "ten-year man."

This was an ideal arrangement for Saxby as it allowed a married man of mature years to achieve a degree, and, at forty-five, Saxby was definitely of maturer years than the typical Cambridge student—his own children were the same age as most of his peers—with only a minimal amount of time in residence at the university. He could continue living and working at Bonchurch while studying. In practice, by the time Saxby enrolled, the academic exercises had become a mere formality, and the whole anachronistic scheme had fallen into disrepute. It had become a kind of loophole through which a man could squeeze his way to a degree with minimal effort, and within another few years it was abolished.

There is no record to show that Saxby served his residence at Cambridge, or even if he took the exams for the divinity degree. From a modern perspective it may seem an odd choice of degree for a man of Saxby's scientific interests, but it is not at all uncommon to find that even quite distinguished Victorian scientists were also clergymen, a profession that usually left a man with plenty of spare time to devote to other interests. If Saxby was simply keen to gain some sort of academic standing, at his age it was probably the only way he could do so. He never used the initials BD after his name, so perhaps he did not finish his degree, but he certainly did make use of his university connection, and often signed himself as "S. M. Saxby, Late of Caius College, Cambridge." In view of the appointment as an instructor that Saxby would eventually take up with the Royal Navy, it was entirely appropriate; a Chaplain and a Naval Instructor were very often one and the same man.

Shortly after his induction into Caius College, in 1845, Saxby made his first foray into the public arena, marking the start of his efforts to establish credibility as a scientist. That year the

British Association for the Advancement of Science (the BA) held its annual meeting in Cambridge, and Saxby delivered a brief communication on Earth's magnetic variation—the difference between true north and north as indicated by a magnetic compass—in which he gave his opinion that there was a correspondence between this variation and the world's mountain ranges. The presentation was not an earth-shattering moment, and if the tone and brevity of the report that appeared in the association's journal are anything to go by, it was met with indifference by his audience. But Saxby was now published for the first time, and his topic, magnetism, would bring him back to the BA a few years later.

∾∾∾

Anyone wandering into the lobby of Washington's Smithsonian Institution in 1850 could hardly fail to notice an addition to the decor. A large map had been placed on one wall, displaying the weather conditions for the previous day in various American cities. A coloured card was placed at each location—white for clear skies, grey for cloudy, and black for rain. Later, symbols were added to indicate wind direction.

How was this possible?

Despite Redfield and Espy's bitter dispute, there was one point on which they agreed. Both men had been quick to recognize that Samuel Morse's electric telegraph, which by 1847 connected Washington, Baltimore, and Boston and was still growing, offered a network that could be used to provide warnings of advancing storms to the seaports on the Atlantic coast, and they each proposed that such a system should be organized. For the first time, here was a means of sending news of approaching bad weather to other locations that lay in its path, faster than the system was moving.

But not everyone was convinced. Strange as it may seem, when meteorologists in the U.S. (and later in the UK) started collecting weather observations systematically, at first they had no intention of using them to foretell the weather. They were hoping only to increase our theoretical understanding of weather systems.

The physicist Joseph Henry, one of the best-known American scientists of his day, was the brains behind this innovation. But it would not have been possible without the extraordinary bequest of an Englishman, the illegitimate son of the first Duke of Northumberland, John Smithson. Smithson had never visited the United States, but for reasons known only to himself he willed his estate, worth more than half a million dollars, for the building of an institution in Washington, DC, whose sole purpose was the "increase and diffusion" of knowledge. The bequest was a stroke of good fortune for America, and gave birth to the estimable Smithsonian Institution.

It was also America's good fortune that Henry was the man chosen to head the brand new Smithsonian, for he was not only one of the country's leading scientists, but also had a keen interest in meteorology. Henry was above all a "pure" scientist—one who was interested in science for its own sake rather than for any practical application it might have. Indeed, he missed at least two opportunities for benefitting financially from his discoveries. He had put together an electric telegraph in his laboratory ten years before Morse, but used it as a novelty to demonstrate to his students. He also discovered electromagnetic induction, the principle that gives us electric motors, alternators, and a load of other gizmos, at the same time as the Englishman Michael Faraday, but Faraday published his results first and won all the acclaim. Henry took home the runner-up prize: the unit of induction was named for him—the henry.

To Henry's way of thinking now, the electric telegraph offered a means for solving the questions that still surrounded the

American Storm Controversy. He knew that if scientists were ever to fully understand the workings of the weather, they would need far more data than they had. But he was also down-to-earth enough to foresee the potential of the system for weather warnings. Like Redfield, Espy, and others, Henry also knew that, since most storms travel from west to east across the United States, the weather telegraph would be a means of providing warnings of their approach to the eastern states. And by the late 1840s, telegraph operators were already sharing local weather information among themselves.

And so, as Secretary of the Smithsonian, Henry suggested to the institution's board that a network of telegraph operators should be organized around the country to transmit daily reports of the weather at their location. The board approved Henry's proposal in December 1847, and by 1849 a system was in place. The telegraph was extremely expensive to use, and the Smithsonian made arrangements with the telegraph companies to transmit weather information only at certain (less expensive) times of the day. At first the network comprised some twenty telegraph operators across the United States, who sent instant and simultaneous daily weather reports to Washington. Henry's staff then transferred the weather data to the wall map in the institution's lobby.

The map was a great hit with the public. It did not attempt to interpret the forthcoming weather, but it was easy to follow the passage of major storms across the country from east to west on the map, from which anyone with a little weather knowledge could take a stab at working out what to anticipate over the next day or so. Within a few years (from 1857 onward), the weather reports were also published in the *Washington Evening Star*, and other newspapers. It was not yet weather forecasting, but Joseph Henry had set the groundwork in place.

In England, Stephen Saxby was beginning to form his own ideas about the underlying causes of weather disturbances, and like so many other practical men who took an interest in the weather, he was motivated by the terrible loss of life and property every time there was a gale near the coast. There was a fearful one in or about the year 1849 on the south shore of the Isle of Wight, near his home at Bonchurch, that wrecked many fishing boats and ruined much fishing gear. Although the boats were stored high above the high-water mark they were swept away by a tremendous tide. Fortunately the Shipwrecked Mariners' Society—an institution in which Saxby showed a great interest throughout his life—was able to help one poor fisherman with a large family, who had lost his livelihood.

Saxby was certainly thinking about the hazards faced by seamen at this time, because soon after this event his first article appeared in the *Nautical Magazine*, a monthly periodical of general interest to seafarers. It was the start of a stream of articles that would appear over the next few years, concerning various ways to improve safety at sea.

The article was titled "Which is the Best Lifeboat?" and in it Saxby pointed out a number of flaws in the design for a new type of lifeboat. The National Institution for Preservation of Life From Shipwreck, as it was initially known (later it became the Royal National Lifeboat Institution), was founded in March 1824, but by 1848 many of its lifeboats were in poor repair. There was no standard design, and often the boats in use were not suitable for the dangerous work—they were too heavily built, easily swamped in heavy seas, and liable to capsize.

In 1851, George Percy, the fifth Duke of Northumberland, became president of the lifeboat institution, and offered a hundred-guinea prize for the best model of a lifeboat that was self-righting and self-draining, with another hundred to build a full-size boat based on the winning model. The competition received 280 entries from Britain, Europe, and the United States, and the prize was won by Mr. James Beeching of Great Yarmouth, England, a boat-builder whose past experience included, among other things, building several boats for use in the smuggling trade—so likely he knew a fair amount about small-boat stability in surf! His design was modified slightly by a member of the institution's committee, and the modern lifeboat was born.

In his article Saxby argued that the Northumberland committee had not paid enough attention to the need of lifeboats to be self-righting in event of capsize, nor to the need for them to free themselves of water quickly. He made twenty suggestions of features that should be incorporated into the design of a lifeboat.

And it was at this point that he made his first public foray into the great debate about weather. In his next article for the *Nautical Magazine*, called "Terrestrial Magnetism and the Law of Storms," which appeared the following year, he expressed some rather bizarre notions about the weather—setting a trend that would persist throughout his career. "The circumstances of cyclones always striking the earth's surface from the eastward," he wrote, "would seem to indicate that thermo-electricity is an active agent in the Law of Storms," and he urged the investigation of the role of the earth's magnetism in storm formation. But, although he did continue to believe that electricity was the underlying cause of storms, it was not an idea that he would pursue with any zeal.

⌣⌣⌣

Sometime in the early 1850s, Saxby left his position at Mountfield School, where he had been teaching young men of school age, and moved from Bonchurch to Rock Ferry, Birkenhead, on the banks of the River Mersey across from Liverpool. He was still teaching, but now he was making his living as a private adult tutor, perhaps preparing young men for university and other forms of higher education, and apparently doing very well. He is said to have been earning more than £2,000 per year, a healthy sum in those days.

It may simply be a coincidence that the training ship *Akbar*, a fifty-five-gun frigate originally brought to Liverpool in 1829 for use as a quarantine vessel, was also moored off Rock Ferry. The ship had lain idle in the Mersey for many years and the Admiralty had loaned it to the Liverpool Juvenile Reformatory Association. The school was intended to reform boys who had fallen foul of the law, and was run on the lines of a naval man-o'-war. It also served another purpose, being viewed by Liverpool shipowners as a ready source of partially trained young apprentice seamen for the merchant navy. There is no evidence that Saxby taught on the *Akbar*, which had its own schoolmaster, but navigation—in which Saxby was an expert—was one of the subjects the boys were taught. It is also quite a coincidence that the ship's commander, William Fenwick, later wrote a testimonial for Saxby. On the other hand, the attitude that Saxby later exhibited toward teaching "factory boys" suggests he may have felt that such a task was beneath his dignity.

Saxby was now hard at work developing a number of inventions, all of which continued to reflect his concern to improve safety at sea. His patents taken out at this time, for instance, include improvements in the gear used for lowering ships' boats, and for holding and letting go tackle—based on what Saxby had learned from studying a serious accident involving poorly designed equipment. Saxby was also getting a taste of foul weather at sea—if he needed any reminder—for in 1853 he was caught in a fearful gale (he called it a "hurricane") aboard the iron-hulled paddle steamer *Ravensbourne* for about thirty-five hours in the North Sea.

Although many of the shipping disasters of the mid-nineteenth century were the result of ships putting to sea directly into the teeth of such storms of which their captains had no foreknowledge, others were due to the appalling condition of the world's merchant fleets—leaky, poorly built, and overloaded ships. And now that iron-hulled ships were rapidly replacing traditional wood-built vessels, navigators were faced with a new problem. The vast amounts of iron used in ship construction played havoc with ships' magnetic compasses. This was brought home in a big way on January 21, 1854, with the wreck of *Tayleur*.

The *Tayleur* was an iron-hulled, full-rigged clipper, on its maiden voyage, bound from Liverpool for Australia, with 528 passengers and crew. On the second day out the ship experienced heavy weather, and on the third the captain realized that his three compasses were all giving different readings. The compasses had been "adjusted" using the conventional techniques of the time, techniques that fell far short of the accuracy, reliability, and consistency required for safe navigation.

The captain believed he was sailing south down the middle of the Irish Channel, when suddenly the steep rocky cliffs of Lambay Island, just off the east coast of Ireland, loomed ahead. It was too late to turn the ship around. In a futile attempt to save it from driving ashore he dropped the anchors, but their cables snapped and the *Tayleur* struck broadside on to the rocks and sank. In all, some 290 lives were lost. Two subsequent investigations both blamed the ship's compasses. A report by the Marine Board of Liverpool declared that "the *Tayleur* was brought into the dangerous position in which the wreck took place through the deviation of the compasses, the cause of which [the Marine Board] had been unable to determine." The board also pointed out that this was just one of many cases in which ships' compasses had proved greatly in error (on wooden- and iron-hulled ships alike). Everyone, it seems, except those who had set themselves up as experts in compass adjustment (including the Astronomer Royal), agreed that the current system of adjusting compasses was dangerously and irresponsibly inadequate.

Stephen Saxby now turned his attention to the problem and prepared a paper that he proposed to present at the twenty-fourth annual meeting of the British Association, which, as luck would have it, in 1854 was to be held in Liverpool. Unlike his first appearance nine years earlier, this time, with such an important issue, he had every reason to think he was about to make a bit of a splash.

CHAPTER 3

Storm Warnings

Those learned men who are honest and careful of their reputation, will never venture to predict the weather, no matter what may be the progress of science.

—François Arago, 1846

By the mid-nineteenth century, the science of meteorology was still poorly understood, people still held many misconceptions about weather, and meteorologists continued to argue over the nature of storms. But the work of Redfield, Espy, Henry, and others in America had shown that the weather is not as bewildering as it seems—conforming instead to patterns that make sense if they are observed over large areas—and that the electric telegraph could be used to provide warnings of approaching storms.

On the other side of the Atlantic was a man who shared this vision. He was to become the pioneering figure of British weather forecasting in the nineteenth century—he actually invented the term *forecasting*—Vice-Admiral Robert FitzRoy.

FitzRoy is best known today as the man who captained HMS *Beagle* during its famous five-year surveying voyage to South

America and the Galapagos Islands between 1831 and 1836. He had suggested that a naturalist accompany the expedition, and the man chosen was the young Charles Darwin, whose theory of evolution was inspired by what he saw during the voyage. FitzRoy had been appointed to the *Beagle* command after its previous captain had put a bullet through his own head, and there is a tragic irony in this, as will become clear later.

Like so many seamen whose lives were ruled by and too often lost to the weather, FitzRoy had a long-standing and profound interest in the subject. He had clearly given it a great deal of serious thought, when in 1843 he appeared before a parliamentary select committee to propose a network of barometric recording stations around the British coast that would provide storm warnings to mariners. Nothing came of the proposal at that time, though, and shortly afterwards FitzRoy sailed for New Zealand, where he had been appointed governor. His thoughts on weather were put on hold for a while.

Another matter of profound interest to FitzRoy was the well-being of his fellow men, and on his arrival in New Zealand he was appalled at the way his countrymen treated the native peoples. As governor, his attempts to uphold their treaty rights against the interests of the white settlers led to his early recall to London. Back in England he returned to naval work, in 1848 commissioning and taking command of a new frigate, HMS *Arrogant*, which was powered by both sail and steam. The task of commissioning took two years and wore him out physically. This, and worries over his personal finances, sent FitzRoy into a dark depression, one of many that he suffered throughout his life. He resigned his command, perhaps thinking that when he was well again there would be the opportunity of another ship; but when he recovered his health, there was no such offer.

The outlook for FitzRoy was bleak at this point, and although he briefly dipped his toes in the commercial possibilities of steam propulsion, becoming a managing director of the General Screw Steam Shipping Company, it was not to last. Fortunately, though, events were about to unfold in the United States that would determine the course of the remainder of his life.

The Redfield-Espy dispute about the nature of storms had finally settled into an uneasy truce, although it was far from resolved. Joseph Henry recognized that both men had made major contributions to our understanding of atmospheric motions, and believed, correctly, that their ideas were not incompatible. But the world of meteorology was not about to remain calm for long. Henry's weather network had been up and running for only a few years when a new feud erupted. This time it was not a disagreement over the science of weather, it was a turf war between Henry and Matthew Maury.

While Henry had been organizing his network of land-based weather observers across the United States, Maury had been enlarging his network of ships at sea, gathering weather data that they forwarded to him in Washington. He was now receiving data from more than a thousand vessels. Maury was also interested in the possibility of weather prediction, and like his contemporaries, he saw the telegraph as the means of making it happen. But if it was to become a reality, he realized, weather observations would need to conform to a universal standard. Maury came up with the idea of an international conference to try and agree on such a standard.

But Maury hadn't anticipated that his proposal for universal standards, on land as well as at sea, would ruffle the feathers of his peers. Henry, already cooperating with the Royal Engineers in Britain to make uniform meteorological land observations, was incensed by Maury's proposal, perceiving it as an attempt to steal his thunder—so to speak.

Robert FitzRoy as a young man.

Henry enlisted the help of his friend Alexander Dallas Bache, the great-grandson of Benjamin Franklin. Bache was superintendent of the U.S. Coast Survey and had his own beef with Maury, whose work on ocean currents he considered an intrusion on *his* territory. Determined to halt Maury—this pretender to knowledge whom they did not regard as part of the professional, university-educated scientific elite to which they belonged, and whose appointment as head of the National Observatory they had long resented—Henry and Bache missed no opportunity to express their opposition to the proposed conference. Their efforts to derail the conference failed, but they did manage to thwart Maury's proposal to have land observations included in the discussions. In the end, in late August and early September 1853, Maury succeeded in bringing together weather specialists from a number of maritime nations to a historic conference in Brussels.

The conference was a resounding success. The British government, whose cooperation Maury wanted more than that of any other nation, had initially shown little interest in the conference, but finally buckled and decided there was merit in his scheme after all and sent two delegates. In all, twelve delegates from ten countries—Belgium, Britain, Denmark, France, the Netherlands, Norway, Portugal, Russia, Sweden, and the United States—met and settled on a standard system of making, recording, and sharing weather observations at sea. It's no coincidence that they were mostly naval men. Even after the conference, a number of maritime nations that had not taken part were keen to sign on to the agreement. Over the next fifty years, American, British, Dutch, and German seamen alone filled out and returned a staggering 26.5 million "abstract logs," the standard form that Maury had devised for them in which to record their weather observations.

Soon after the conference, the British Government announced plans to set up a Meteorological Department (later to become the Meteorological Office), under the control of the Board of Trade, and funding was approved in the House of Commons in June 1854. There was a telling moment in the House when one MP suggested, quite seriously, but overestimating the perspicacity of his peers, that the department's weather observations on land and sea might eventually provide a picture of the weather in London twenty-four hours ahead. His comment was met with laughter from the backbenches. Such was the prevailing attitude toward foretelling the weather.

The Royal Society recommended Robert FitzRoy for the task of running the new department, and he took up his new post in August 1854. His official job title was "Meteorological Statist," reflecting the nineteenth-century term for a statistician, and he was expected to gather and sort weather data collected by ships at sea, much in the way that Maury had been doing in the United States. The wise men of the Society believed this was the path to the understanding of weather systems.

But his employers had misjudged FitzRoy. They could not have picked a man less suited for the role of a well-paid filing clerk, and FitzRoy had far greater aspirations in his new job. He interpreted his orders to mean much more than the mere gathering of weather statistics. What is the point of assembling large amounts of data if you are not going to do anything with them, particularly when there are practical benefits to be gained from all the data you are collecting? Robert FitzRoy also knew from his own experience the devastating damage that storms can inflict on shipping, and he had long hoped for some means of issuing storm warnings for sailors. The loss of shipping and lives in British waters had reached staggering proportions. Between 1852 and 1855, an average of almost a thousand ships and more than nine hundred lives were lost around the coast each year. Besides, part of FitzRoy's mandate was to enable mariners to use their vessels more efficiently.

Within three months of FitzRoy's appointment, there was a maritime disaster that brought the issue even more sharply into focus. In 1853, Britain and France had declared war on Russia—the Crimean War—and on November 13 of the following year, when the allied battle fleets were assembled in the Black Sea off the Crimean peninsula, they were devastated by a massive storm. Many ships dragged their anchors or broke their anchor cables and were driven ashore. More than thirty French and British ships, mostly transports, were wrecked, taking with them the lives of some 340 men. Others had to jettison their guns to lighten ship, and many more were dismasted. Much of the army's supplies and ammunition went down with the ships, along with winter clothing for the troops. It was a crippling loss.

Before demolishing the battle fleet, the storm had passed across the European countries to the west, and it wasn't long before people started to wonder if forewarning of the storm could have been provided using the electric telegraph, as had been done in the United States.

* * *

The twenty-fourth annual meeting of the British Association for the Advancement of Science, in Liverpool, was a grand affair. It was held in late September 1854, in the glittering new St. George's Hall, an ornate neoclassical edifice, housing concert halls and law courts, on the port city's famous Lime Street. The plaster on the walls of the hall was barely dry.

The meeting opened with a report on Matthew Maury's scheme for the improvement of navigation and a plan put forward by Robert FitzRoy to implement a standard system of recording weather at sea, using a slightly modified version of Maury's system.

Like all such meetings, the main business was the reading of a large number of scientific papers to the Association's seven sections, each devoted to a different branch of science. In addition there were soirées, *conversaziones*, general meetings, evening excursions, and a couple of major discourses on subjects of special interest. One of the papers on the agenda, scheduled for the last day of the meeting, was "Mechanical Appliances on Board Merchant Ships," and it was to be read to the Mechanical Science Section by Mr. Stephen Saxby.

Magnetism in iron hulls had become such a pressing issue since the *Tayleur* disaster that the BA meeting devoted considerable time to its discussion, and a number of eminent scientists were addressing the issue at Liverpool. William Scoresby, the famous oceanographer, presented a paper on the loss of the *Tayleur*. Scoresby brought forward a number of ideas about the cause of compass deviation—the error in a magnetic compass reading induced by the iron in a ship's hull—and gave demonstrations using iron plates. And one of the meeting's major discourses, on the earth's magnetism, was delivered by Colonel Edward Sabine, a leading expert on the subject who in a few years' time (in 1861) was to become president of the Royal Society.

On the day that Saxby was due to deliver his paper, John Scott Russell, an engineer and naval architect who had worked on ship design with the brilliant engineer Isambard Kingdom Brunel, subjected the gathering to a marathon two-and-a-quarter hour address on the topic of magnetism and ships. Even the BA's Mechanical Science section could endure only so much on magnetism, and when Russell finally sat down the organizing committee approached Saxby and asked him if, since there had already been much discussion on the subject of his address, he would be prepared to confine his talk to explaining models of his several inventions. "Although flattered with the importance the Committee placed on my patents," Saxby later remarked, "I was not anxious to trouble the meeting with my private matters, and begged therefore to withdraw."

It must have been a bitter disappointment for Saxby. He had independently come to much the same conclusions as the highly regarded Scoresby on the problems of compasses in iron ships on their maiden voyage, even if his explanations differed somewhat. Although he says he withdrew of his own volition, perhaps he was merely putting on a brave face. It's far more likely that he was politely requested to step aside because his paper duplicated much of William Scoresby's.

One of the outcomes of the BA meeting was the formation of the Liverpool Compass Committee to address the issue of compass deviation, and to present its results to Parliament. It was an issue that continued to prey on Saxby's mind too, but there was another aspect of the *Tayleur* incident he felt he had not yet resolved—the problem of ship's anchor cables breaking when put under extreme strain. When a ship's anchor cable runs out rapidly and is suddenly snubbed, as often happens when the anchor is let go in an emergency, the cable tends to snap, with dangerous—lethal in the case of the *Tayleur*—consequences for all on board. Saxby was soon back at his drawing board in Rock Ferry, and the solution he came up with was simple and inexpensive: a device designed to prevent chain running out too fast, which became known as "Saxby's Stopper."

Saxby also started to think about another device that would help navigators get around the problems of compass deviation, and the following year he patented one of his most innovative creations, the "Spherograph." It was an aid to navigation, a simple labour-saving device, consisting of two circular cards, the upper one transparent, attached by a pin through their common centre so that they could rotate one over the other. On each card were drawn a number of circles, as well as parallels of latitude and meridians of longitude. It could be used by mariners to solve a number of mathematical problems in navigation, and, in the wake of the *Tayleur* disaster, Saxby also promoted its use in "ascertaining the errors of mariners' compasses," presumably by being able to quickly establish the direction of the ship's head and compare it with the compass reading to determine how much the compass is in error. His friend William Fenwick, retired commander of HMS *Rollo* and now in command of the training ship HMS *Akbar*, described it as one of the best systems of navigation ever invented. The device became commercially available the following year, but it was not the immediate success that Saxby had hoped. Apart from a couple of encyclopedia articles about it, almost certainly written by Saxby himself—he contributed all the articles on nautical matters published in the *English Cyclopedia*—there is little mention of it to be found.

At the same time, Saxby also published a number of suggestions in the *Nautical Magazine* on how ships might avoid collisions at sea. Long before there were laws regarding the "rule of the road" at sea, seamen had evolved a system for avoiding other ships. As far back as 1645, the Earl of Warwick had declared in a set of sailing instructions that no captain shall take the wind of an admiral. Perish the thought! By 1846, informal rules that had become part of the custom of the sea were replaced in British waters by regulations that carried the weight of the law concerning which ship should give way to the other when two ships meet, depending on the circumstances. They were far from perfect, however, and Saxby had his own thoughts on how ships might signal each other to communicate their intentions. By day he suggested the use of hand signals, and at night the use of a "danger light" in addition to the regular lights carried by all vessels, whether under steam or sail. Although he was aiming for simplicity, Saxby's ideas were totally impractical. They were not taken up the authorities.

Saxby also addressed another, more benign hazard faced by ships' captains: the presence of fashionably dressed lady passengers. He describes a steamship captain in the Royal Mail Service who had noticed that the presence of a lady near his binnacle affected the compass reading. The captain eventually realized the effect was due to her iron-framed chair and the steel hoops within her ample crinoline!

Another matter was starting to occupy the mind of Stephen Saxby while he was living at Rock Ferry. If he had not started doing so earlier, he was by now observing and keeping careful records of the weather. And he thought he was starting to detect a remarkable pattern. The timing of storms seemed to coincide, he thought, with certain stages of the moon's orbit.

◦◦◦

Matthew Maury had accumulated a wealth of information from his studies of the oceans and the atmosphere, and had developed his own theories on their circulation. He was now preparing to publish them. At first he intended to include his ideas in a new edition of the directions that accompanied his *Wind and Current Charts*. His publisher, however, thought this new information was interesting enough to appeal to a wider audience, and the chapter rapidly grew into a book, *The Physical Geography of the Sea*, published in 1855. Maury certainly had a gift with words. "There is a river in the ocean," the book's evocative opening lines, have become forever linked with the Gulf Stream. An immediate success with the public, *Physical Geography* was not well received by Maury's critics in the scientific world, however, who slammed it for its many inaccuracies, and were irritated by the many religious references that peppered his prose.

Shortly after Maury published his book, a brilliant, mostly self-taught thirty-seven-year-old teacher stepped into a Nashville bookstore and purchased a copy. The teacher's name was William Ferrel, and that purchase marks a turning point in meteorology. The book aroused Ferrel's interest in the subject, and when he applied his powerful intellect to its unresolved questions he made another leap forward in our understanding of storms. And this, in turn, finally laid the Redfield-Espy dispute to rest.

Ferrel came from a poor background as a farm boy in Bedford County, Pennsylvania, and moved to West Virginia when he was twelve. His schooling was, to say the least, second-rate, but he was an intensely curious young man, and intelligent enough to educate himself. By his early twenties he was making a living teaching school, and devoting much of his spare time to a study of science. He was interested in, among other things, the action of the sun and moon upon the tides, and by the time he had reached his early thirties he had written a scientific paper in which he proposed that the action of the sun and moon had a tendency to slow the earth's rotation. That's not bad going for a man who had spent so much of his youth wielding a pitchfork that he had little time to study—he always regretted his youth as having been wasted.

Ferrel realized that much of Maury's book was deeply flawed. He was asked to write a critical review of the book but declined. A chronically shy man, he could barely bring himself to read his own scientific papers before an audience, and he disliked any kind of controversy. Instead, he decided to write another scientific paper in which he put forward his own ideas. It was called "An Essay on the Winds and Currents of the Ocean," and was published in 1856 in a most unlikely sounding venue, the *Nashville Journal of Medicine and Surgery.*

He followed it up with other papers that cemented his ideas, and in one of them he put forward the important suggestion, which became known as Ferrel's law, that "If a body is moving in any direction, there is a force, arising from the earth's rotation, which always deflects it to the right in the northern hemisphere, and to the left in the southern." The bodies he was thinking of were the air molecules that make up the atmosphere. Does that sound familiar? Remember the Coriolis effect, mentioned in the previous chapter? What Ferrel had done was show that the Coriolis effect—neglected by atmospheric scientists for a quarter of a century—works in the same way on the air particles in the atmosphere as it does on cannonballs.

Ferrel's law explained why Redfield's storm maps actually showed winds moving—not around a storm centre in concentric circles, but slightly inward toward the centre. Redfield had believed this was simply due to errors and inaccuracies in measuring wind direction. Now it was clear that the winds within a storm start off by blowing toward the centre, but are deflected so they end up *almost* revolving around the centre, but *not quite.* Ferrel also agreed with Espy's theory that heating of the atmosphere, driving water vapour upwards until it condenses, was the force that sets hurricanes in motion.

Ferrel has been described as the first person to really understand the mechanics of the atmosphere, and he remains one of the most important figures in nineteenth-century weather science. There was one man, however, who would continue to disagree. If you haven't guessed already, that man is James Pollard Espy, and he always refused to budge one inch over his theory that winds blow directly toward the centre of a storm. Nevertheless, as far as everyone else was concerned, Ferrel had finally brought the American Storm Controversy to a conclusion.

~~~

One of the people who believed it would have been possible to warn the Anglo-French Crimean fleet of the storm that wrecked so many of its ships was the director of the Paris Observatory, an astronomer called Urbain Le Verrier, the man who discovered the existence of the planet Neptune. After the Crimean War ended in February 1856, Le Verrier collected weather records of the storm from England and Europe, from which he reconstructed its passage. He concluded that a warning of the storm's approach could indeed have been telegraphed to the fleet lying off Balaclava, giving the navies time to prepare. He proposed a European network of observers, like Joseph Henry's in the U.S., to gather and transmit weather data. The French scientific establishment, however, was firmly opposed to any kind of weather prediction. One man in particular, François Arago, Le Verrier's predecessor as director of the observatory and the man who famously made the remark quoted at the beginning of this chapter, resisted Le Verrier's efforts. It was several years before the French government established a telegraphic storm warning network, and even then it had a troubled existence under Le Verrier's autocratic rule.

In Utrecht, Holland, at the same time, the founder and director of the Royal Netherlands Meteorological Institute, Christoph Buys Ballot, was quietly going about his business and, although he was not in favour of weather forecasting in general, he beat everyone else in Europe by making his country the first to actually put a system of storm warnings in place in 1860. He has received little credit for this achievement, but is perhaps best known for the practical law that bears his name—Buys Ballot's law—which mariners ever since have found rather useful: if you stand with your back to the wind in the northern hemisphere, the centre of low pressure (think *storm centre*) is located on your left hand side.

In Britain, FitzRoy's Meteorological Department was soon to follow suit.

∾∾∾

By the late 1830s, steamships were providing ferry services around the British Isles and across the Channel to France. British-built steamships were even operating as far away as Australia. In the early 1820s the Royal Navy had commissioned a handful of steam-powered vessels—not fighting ships, but mainly tugs for manoeuvring sailing vessels in harbour, and even for towing men-o'-war into battle in a calm, or when an enemy lay upwind. But eventually, the navy converted all its men-o'-war to steam.

When these new and highly technical steam-powered vessels appeared, the navy needed trained men to operate and maintain them. In the early days of steam (until 1837), the service had relied on civilians, known as "engine-men," supplied by the manufacturers. This worked well enough when there were only a handful of steam-powered vessels, but as the steam fleet grew and the ships became bigger it became less satisfactory, and the navy started recruiting its own engineers. As engines became more sophisticated, there was also a need for engineers with greater technical knowledge, and the Admiralty started offering higher salaries to attract men who were already trained and experienced. But, try as it might, it could never find enough trained engineers at this period to man the rapidly growing steam fleet. It could not rely on the merchant fleet for recruits, as only a very small proportion of merchant ships were powered by steam, and generally the pay and the conditions on these ships were better. Clearly there was a need for the navy to train its own engineers, and a need for instructors to do the training.

Unfortunately, for the two decades from 1837 to 1857 the training and education of young officers in Britain's Royal Navy was in a total shambles. The rot started in 1837 with the closure of the Royal Naval College Portsmouth, to cut costs. During the previous century, naval officers mostly learned their profession at sea, under the guidance of sea-going schoolmasters. The system had its faults—in practice many young officers received very little schooling, and the efforts of the schoolmasters were often hindered by established officers—but it persisted because senior officers were in favour of combining study with practical experience in the arts of seamanship, leadership, and sometimes battle. But even as early as the 1730s the Admiralty had realized the benefits of educating young officers in a more formal shore-based setting, and had established the Portsmouth Naval Academy. For the next hundred years sea-going schoolmasters and the academy coexisted, side-by-side. The Academy did close in 1806, but it had been quickly replaced by the Portsmouth College.

Now that the College was gone, after 1837, the Navy reverted to the earlier system of instructing its young officers solely at sea.

The ship's company of every naval man-o'-war was to include a schoolmaster, but unlike the schoolmasters who had shipped aboard naval vessels previously, they were now known as "Naval Instructors." As a sign of their importance (perhaps) instructors were given the equivalent rank of Wardroom Warrant Officers. (The college did reopen the following year, not for new recruits, but as a college of higher education. Later, in 1853, it did offer courses for engineers, but then its main focus was on training for existing executive officers, not on training the new breed of engineers.)

Although the navy had hoped to recruit university-educated men as instructors, at first it attracted only a handful. Then the number of graduates increased, but only because of a rather peculiar naval regulation that offered an extra incentive to naval chaplains already in the service. If they took on the additional role of instructor they received three-quarters of an instructor's full pay as well as a £5 levy for each student taught, on top of their chaplain's wages—a reasonably cheap and convenient solution for the Admiralty. As a result, there was a rush of clergymen to fill the available instructor positions. By 1846 almost half the ninety-six naval instructors were men of the cloth.

This had the unfortunate effect of providing a disincentive for non-clerical graduates to apply for (and remain in) positions as instructors—they would receive less pay than chaplain-instructors for doing more or less the same job. The navy also appointed chaplains by *commission*, while instructors were appointed by *warrant*, a considerable difference in status. In an effort to attract enough competent instructors for its ships, the navy increased their rate of pay. At the same time, the number of engineer officers was increasing dramatically, almost doubling between 1850 and 1857 from 440 to 862.

The navy finally addressed the crisis of educating its executive officers in 1854, by bringing the first officers' training ship, an elderly man-o'-war, HMS *Illustrious,* into operation, complete with a teaching staff, comprising (inevitably) two chaplains and a naval instructor. A second instructor was soon added. Cadets spent a year training aboard this harbour-bound vessel before transferring to a sea-going training ship. By 1859 a bigger ship was needed and *Illustrious* was replaced by HMS *Britannia.* In addition to their regular studies the cadets also received occasional lectures on the new method of propulsion—steam—from a naval engineer.

As for the engineers themselves, the navy maintained a half-hearted attitude toward their training. It continued to expect to hire men who were already trained, or to promote apprentices and craftsmen from the naval dockyards, for whom schools were set up in the principal dockyards to provide a basic education. Woolwich dockyard, with its steam factory where engines were built for the navy's ships, was the first, but these schools were never intended to produce the much-needed engineering officers who would oversee the increasing numbers of engine-room hands needed to operate the new machinery.

It was in the Royal Naval Dockyard at Sheerness that the navy finally decided to set up another new type of training school, specifically for the education of its engineer officers. A large number of such officers were stationed at the port, and they had plenty of leisure time for study. The school was to be based aboard HMS *Devonshire*—a third-rate, seventy-four-gun warship launched in 1812. After an unremarkable career, the ship had been put into harbour service in Sheerness in 1849, where it was used as a depot for Russian POWs during the Crimean War. *Devonshire* was now one of several hulks lying idle ("in ordinary," in the naval jargon) at the dockyard, where a number of steamers not in commission were also mothballed as part of Her Majesty's Steam Reserve.

The Admiralty needed a qualified man to take charge of this unique new floating school, and first approached a Cambridge

University fellow to see if he was interested. The man knew nothing of nautical matters, however—but he knew of someone who did fit the bill. Stephen Saxby's earlier life at sea with the East India Company, his subsequent career as a tutor, and his scientific work made him an ideal candidate for the position of headmaster. Saxby applied for the position and visited Sheerness in January 1858, where he met the Captain of the Reserve, Captain Halsted, who was stationed aboard HMS *Cressy*, the Reserve's Guard Ship that oversees the dockyard's marine affairs. It was Halsted who had first proposed the appointment of a headmaster, and he gave Saxby a briefing on the nature of the proposed school.

In his application Saxby listed thirty years teaching experience in every branch of classical and English study, as well as experience in demonstrating mathematics, mechanics, steam, and other technical subjects, and lecturing on chemistry, nautical astronomy, and general science. He also professed expertise as a marine and land surveyor and a draughtsman, and a thorough acquaintance with both the theory and practice in all matters nautical. "I am known to most of the older officers of the hydrography department at the Admiralty; to the Trinity Board; to Captain Walker HEIC (Naval head of Board of Trade)." Along with this he included two testimonials concerning his teaching ability and aptitude from two clergyman who had long resided at Bonchurch, Isle of Wight, and a third from retired naval commander William Fenwick in Liverpool.

Normally candidates for the role of Naval Instructor were required to be between the ages of twenty and thirty-five and had to undergo a rigorous examination process. They were also expected to have a wide area of knowledge and to show proficiency in arithmetic, algebra, plane and spherical trigonometry, mechanics, hydrostatics, French, and the classics. Then they had to pass a final examination in navigation and nautical astronomy and other related subjects, and to show proficiency in making

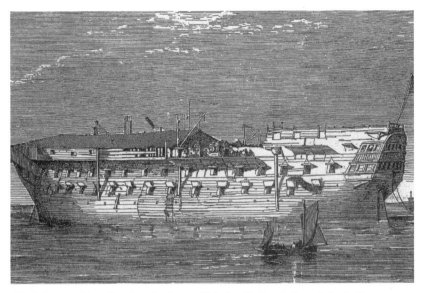

HMS Devonshire. *Stephen Saxby and his family lived aboard the hulk at Sheerness from 1858 to 1865.*

observations with the sextant and using the azimuth compass, chronometers, and other nautical instruments.

Perhaps because it couldn't find a suitably qualified man within its own ranks, or perhaps because the new establishment was something of an experiment, the Royal Navy had been forced to look further afield for a candidate to take charge of its new school; and, either way, the navy had to bend the rules to take on Stephen Saxby. He was now fifty-four years old—almost twenty years older than allowed under the regulations. And he was not required to sit for an exam. The navy seems to have got around these irregularities by appointing Saxby an "acting" Naval Instructor, a caveat that eventually cost Saxby dearly, as it meant the position offered no real security. He was exempted from examination because of his impressive credentials and because, in the navy's words, *he might at any time be removed*, despite the fact that Captain Halsted had understood the new position under his command was to be permanent, subject only to satisfactory performance.

Saxby took up his appointment on May 21, 1858, and was officially posted to HMS *Cressy*. The Navy Lists show that he was assigned to a number of different ships belonging to the Steam Reserve at Sheerness during his appointment over the next few years, but this was a formality. In practice he was provided with accommodation for himself and his family aboard *Devonshire*, where he began signing himself as Chief or Principal Instructor.

And so Saxby began an eleven-year stint with the Royal Navy as head of a prestigious naval institution. It must have seemed like a fine way to culminate his teaching career, a job for which he was eminently qualified. But it was not without sacrifice. Saxby gave up his private tutoring and relinquished the rights to his patents, worth some £20 a month, and joined the service at what he regarded as a quite inadequate salary. He quit a comfortable life ashore for a life afloat, moving his family aboard the hulk *Devonshire* amidst the bleak and often foggy marshes of the Thames Estuary. His four older children were now all in their twenties and most likely had left home already, but the two youngest, fourteen and seventeen, were almost certainly still living with their parents.

Saxby and his family found they had been given spacious cabins in the officers' quarters on *Devonshire*—the whole of the poop deck and part of the main deck—but the navy had neglected to provide them with any furniture. Saxby complained, and the navy promptly approved furniture for his cabins, "as for a second-rate ship of the line," along with glassware, china, linens, and so on. They also allowed him a servant, one shilling a day for lights, and the same in lieu of provisions, along with a four-oared gig and four hands, solely for his own use. It was about the standard of living that a man in command of the ship could expect. It was a promising start to his new career.

Stephen Saxby's new home, Sheerness, was one of Britain's

smaller naval establishments, not on the scale of others like Portsmouth or Devonport, which were also used by the Steam Reserve. Most of the ships built here were sloops or gunboats, with only the occasional larger vessel. In times of war the dockyard's location—on the Isle of Sheppey where the River Medway flows into the Thames at its mouth—made it an ideal station for resupplying and repairing ships, and for this reason the yard had been kept very busy during the Crimean War. It was also the location of a factory for building the navy's new steam engines.

The Romans were the first to recognize the island's strategic location, but it was not fortified until the sixteenth century, by Henry VIII, and the military establishment on it grew piecemeal from then on. It has an unusual, if dubious distinction: captured in the seventeenth century by Dutch raiders, the Isle of Sheppey is the only part of England since the Roman invasion ever to fall— even briefly—into foreign hands.

Sheerness has another dubious distinction. It was the location of the famous Nore Mutiny of 1797. The mutiny followed a similar episode at Spithead (Portsmouth) earlier in the year, when a mob of disgruntled seamen had taken possession of several naval warships and made demands for improved conditions, which had sunk to an all-time low. The grievances concerned poor pay, a lack of shore leave, and the quality and quantity of their food rations. At Spithead the affair resembled a trade union strike more than a mutiny, with its delegations and negotiations, and it ended with the Admiralty's agreeing to most of the men's demands. But when the mutiny spread to the naval ships anchored off the Nore (a sandbank in the Thames Estuary just off Sheerness), the seamen there made further demands over pay and the right of crewmen to have a say in the removal of officers they deemed incompetent, brutal, or addicted to the bottle. At the Nore, however, the Admiralty stood its ground against what it considered unreasonable demands. Eventually the mutiny fizzled, and twenty-nine of the leaders were hanged, which perhaps explains why nineteenth-century naval men sometimes referred to Sheerness as "the last place God made." It was a sentiment that Stephen Saxby might well have echoed ten years later.

∾∾∾

In England in 1859, another terrible tragedy gave urgency to Robert FitzRoy's vision of effective storm warnings. At daybreak on the morning of Wednesday, October 26, 1859, on a hilltop overlooking Dulas Bay in Anglesea, North Wales (near the village of Moelfre), a man was blown out of his bed when the ferocious gale that had howled all night with winds of sixty to one hundred miles per hour from the north-northeast started to strip his roof off. When he and a neighbour climbed a ladder to try and make a temporary repair, he happened to cast a glance seaward. He could hardly believe his eyes, for there below him lay the vast dark shape of a stricken ship that had been blown onto the rocky shore by the terrible storm just before dawn.

The two men rushed to where the ship was being pounded by thunderous waves and raised the alarm. Soon almost thirty local villagers, mostly fishermen and quarrymen, had run to the wreck, where, at great personal risk, they attempted to save the lives of the passengers and crew. One valiant crewman, Joseph Rogers, was able to swim ashore with a line around his waist, but before the unfortunate passengers could be brought ashore on the line a massive wave broke the ship's back and many were thrown into the water. Of the 498 people on board the iron-hulled ship, only thirty-nine were saved. The others were mostly battered to death by the raging seas.

The ship was the *Royal Charter*, owned by the Liverpool & Australian Steam Navigation Company, and it was on passage from Melbourne, Australia, bound for Liverpool with a vast horde of gold in its strongroom, rich pickings from the gold rush in Australia.

The *Royal Charter* was not the only ship to be lost that day. The seas around the coast were driven into a fury. At Liverpool, a pilot vessel was lost with all hands. The screw steamer *Semaphore*, which had left Belfast on Tuesday morning with a cargo of cattle and pigs had lost a number of them overboard in the storm and its decks were a shambles, with dead and dying creatures everywhere. On the other side of England, in Hartlepool, where many colliers and other laden cargo vessels had run for shelter, a great number were driven ashore and damaged. And at Dover the captain of a schooner trying to make harbour misread the lights, missed the entrance and ran ashore. He and two seamen were drowned.

The storm passed directly across central England. Ashore, railways were washed out along the south coast. Sea walls and embankments were breached and esplanades washed away. Several lives were lost, many people were injured by falling buildings, and trees and buildings were damaged (roofs were blown off, chimneys blown over, and windows blown out). In Peckham, London, a woman was blown off her feet by the wind and swept into a canal, where she drowned. The storm coincided with a high tide and there was much coastal flooding. At Eastbourne the bathing machines were carried by a flood into the middle of the town.

But none of the other tragic incidents of that storm could compare with the scale of the *Royal Charter* disaster. A shocked FitzRoy turned his attention to the weather reports that had been gathered during the tempest. The loss was partly the fault of the captain, for another nearby vessel, an American ship called the *William Cumming*, had headed out to sea and suffered no damage whatsoever because its captain, far more weatherwise than the man in command of the *Royal Charter*, had known that toward the open sea he would be moving away from the worst of the storm's wrath. FitzRoy soon concluded that, with the facilities at his disposal, it should have been possible to issue a warning and to avert the disaster.

FitzRoy presented his initial findings on the storm to a meeting of the Royal Society in December 1859, and a few months later he gave another presentation to the BA in Oxford, where he made his case for a storm warning system. This meeting has gone down as one of the most significant in scientific history, for reasons quite unrelated to storms. It was the scene of the classic confrontation between the Bishop of Oxford, Samuel Wilberforce, and Thomas Henry Huxley, the ardent supporter of Charles Darwin, in which Huxley made mincemeat of "Soapy Sam," as the bishop had been nicknamed. But FitzRoy was a deeply religious man, whose Christian views—a literal belief in the words of the Old Testament—were more or less fundamentalist. He was embittered by his past connection with Darwin and he could not avoid expressing his outrage at Darwin's conclusions in an angry outburst. But for FitzRoy all that was a mere sideshow. He won the support he was looking for on the issue of the storm warning system, and the BA Council passed a resolution "praying the Board of Trade to consider the possibility of watching the rise, force and direction of storms and the means for sending, in case of sudden danger, a series of storm warnings along the coast."

Soon after, the board granted FitzRoy formal permission to issue storm warnings to threatened parts of the coast, and he spent a frantic few months organizing his network, initially setting up instruments at thirteen British recording stations, making arrangements with the telegraph companies, and engaging telegraph operators to record and transmit the weather data to London, where if FitzRoy detected the likelihood of a storm anywhere around the British coast he could telegraph a warning to the affected areas. He also devised a simple but ingenious set of visual warning signs to be hoisted at threatened ports to warn mariners if it would be dangerous to put to sea.

On February 6, 1861, Fitzroy's office issued its first storm warning. That day, a storm of Force 10 on the Beaufort scale (winds of more than 48 knots, 89 km/h) was reported at Dover, and Force 9 (winds of more than 41 knots, 76 km/h) at Nairn in Scotland. Elsewhere winds almost reached gale force all along the east coast. A fleet at Shields, in northeast England, ignored the warnings and, as a result, many vessels were wrecked and many sailors lost. In the following weeks FitzRoy issued a number of other warnings, and there was an immediate and dramatic fall in the number of shipwrecks.

Yet only ten days after FitzRoy's first storm warning, his own employer, the Board of Trade, tried to dissociate itself from this aspect of his work. In a simpering letter to FitzRoy, the assistant secretary to the board, T. H. Farrer, a bureaucrat who once quipped that the board was "a sort of wastepaper basket into which matters are thrown which are not wanted elsewhere," and is said to be "the architect of the nineteenth-century Board of Trade," requested that FitzRoy not make the Board of Trade or its head "responsible for anything except the correct relation and translation of facts." He added, "Could [your forecasts] not be managed by making it distinctly appear on the face of what you send out that they do not come from the Board of Trade, but from some other body?"

"Responsibility I never shrink from," was FitzRoy's bitter response, with its scathing implication of Farrer's cravenness—in an otherwise polite letter pointing out that opinions around the coast were entirely in favour of the Board of Trade forecasts.

By a quirk of timing, FitzRoy's first storm warning was issued the same month that the outbreak of the American Civil War brought Joseph Henry's weather program at the Smithsonian Institution to a halt. Henry's observers in the southern states

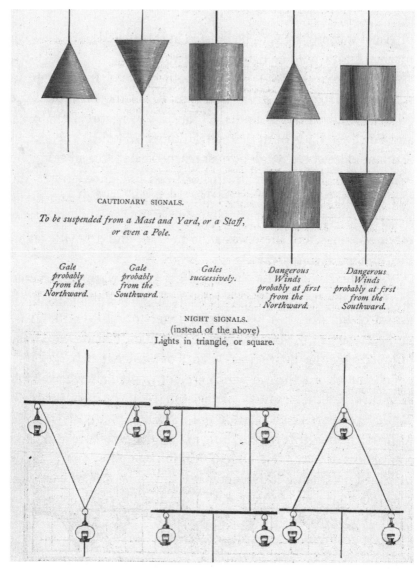

FitzRoy's storm warning signals.

stopped sending their reports, and in the north the military commandeered the telegraph lines for its own use. The outbreak of the Civil War was also a disaster for Matthew Maury. A Virginian, Maury was a Southerner to the core and given his sympathies for the Confederate cause, his post with the United States Navy was no longer tenable. He reluctantly resigned his commission, and went to work instead for the newly formed Confederate Navy.

And FitzRoy's efforts were soon to bring him more grief than he would have thought possible.

CHAPTER 4

Barking at the Moon

Every weather-wizard appeals to the Moon.
—Vice-Admiral Robert FitzRoy

British mariners were quick to realize the value of FitzRoy's storm warnings, even though the warnings had their drawbacks. In the latitudes of Britain (as in North America), there is a large-scale movement of the atmosphere from west to east, so the majority of dangerous storms in British waters arrive from across the vast expanse of the Atlantic Ocean. Not until the invention of the wireless telegraph was there any means of quickly transmitting observations taken at sea to London. Also, FitzRoy based his warnings on observations of existing weather and only issued them when he was sure he knew how the weather might develop and move. He could issue warnings only one day, or at the most two days, ahead of an impending storm, and they were only available to mariners who remained close to the coast, in sight of the storm signals.

Stephen Saxby thought he could overcome this problem of short warning times. Initially he did not think he could do *better* than FitzRoy. He had a great deal of admiration for both the man and his warnings. "Indeed, I do not see how the Admiral's telegram cautions *can* be otherwise than correct generally," he wrote, knowing full well that FitzRoy's warnings were based on solid observations. But, in Saxby's view, their value was limited. The beauty of the system he had developed, based on his belief that he could foretell the weather using certain aspects of the moon's orbiting motion, was that it could offer long-range predictions, so that captains could go to sea and sail well beyond the range of FitzRoy's storm warnings, armed with foreknowledge of when storms were likely to occur. "Why should our signals, or warnings, or forecasts," he railed, "be limited to vessels near shore, when other means exist of avoiding danger? When months, and even years before hand, such warnings may be issued with infallible accuracy?" And for those who were more than an hour or two's sail from land, he offered this: "What if I can supply the warning for those so circumstanced! (I am about to do so)."

Saxby appealed for cooperation among all meteorologists. "No one can contemplate the lamentable sacrifice of human life which every gale causes without an ardent desire to see improvement. We who *think* ourselves well posted up in meteorology, whether with official authority or as lunarists or otherwise, would do well to sink all differences and prejudices for the promotion of the one object."

His lunar method of predicting stormy weather had very little to do with the moon's regular monthly phases—new moon, waxing, full moon, and waning. These he considered largely ineffective. Instead, he looked to another aspect of the moon's rather complex orbit. During its travels around the planet, the moon is always directly overhead somewhere on Earth, and this point, its *terrestrial position*, is always moving. Sometimes it is north of the equator, sometimes south, crossing regularly back and forth. Its distance north or south of the equator is called its *declination* (think "latitude").

The theory was simple enough. Saxby concluded, from what he stated to be a lifetime of weather observation, that at certain times—when the moon's terrestrial position actually crosses the equator, and when it reaches its furthest north or south declination, events that occur about once a week—there would be atmospheric disturbances, or storms. And although he did not think the moon's phases normally had any perceptible effect on weather, when a *new* moon coincided with either of these events, he thought, its influence was heightened considerably.

There was one further aspect of the moon's orbit that Saxby claimed sometimes plays a part in its effect on weather—its distance from the earth. This distance is constantly changing. When closest to the earth the moon is said to be in *perigee*, and when farthest away, in *apogee*. Saxby did not believe that the distance was important generally, with one important exception: when the *new* moon was in *perigee* at the same time as it crossed the equator or reached its maximum declination north or south, these were times that the greatest atmospheric disturbances—the most violent storms—could be expected.

With these conclusions, it was a simple matter for Saxby to reach for his *Nautical Almanac*, extract precise information on the moon's position, phase, and where it stood in the perigee/apogee cycle for any day of the forthcoming year or so and make his predictions many months in advance. He took a few other twists into consideration, but those formed the nub of it.

By now Saxby had abandoned his earlier opinion that there is a connection between the weather and the earth's magnetism, but he did still cling to the popular theory that electricity was the ultimate cause of storms. As the sun and moon vary their positions with respect to the earth, he thought, so does the intensity of their attractions vary. This in turn upsets the electrical charges in the atmosphere. The effect depends on whether the air is positively or negatively charged when the moon effects one of its disturbances,

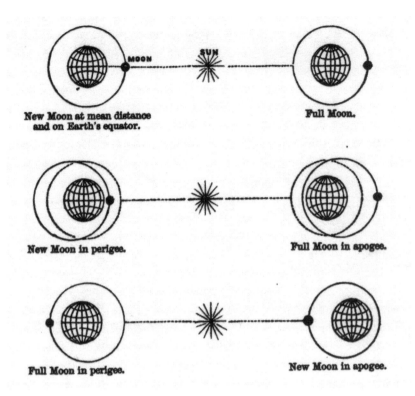

New Moon at mean distance
and on Earth's equator.

Full Moon.

New Moon in perigee.

Full Moon in apogee.

Full Moon in perigee.

New Moon in apogee.

Saxby's illustration of perigee and apogee.

which causes moisture either to condense or to evaporate. A sudden condensation, Saxby argued, causes a partial vacuum into which air rushes, sometimes for several days, until equilibrium is restored. (This explanation was, of course, just as wide of the mark as his previous one.)

The year following the wreck of the *Royal Charter*—a storm that Saxby attributed to the moon's crossing of the equator, and which had also stranded *Devonshire*, demolishing the end of the Sheerness pier—he found in the *Nautical Magazine* an editor who was willing to air his ideas and perhaps test his predictions. He started to pen a series of articles on the moon's influence on the weather, in which he explained his theory. He also gave a number of predictions of when disturbances would occur. At first the magazine had nothing but praise for Saxby, and after his first few articles it ran a glowing editorial:

> *Mr. Saxby appears to be in a fair way to establish a*
> *claim to the grateful thanks of all who are interested*
> *in a foreknowledge of the weather (and who is*
> *not?) by the success which has hitherto attended*
> *his warnings on this subject. He has (as he says) in*
> *our July number…foretold the days on which dirty*
> *weather may be expected in the remaining months of*
> *the year. Our seamen and fishermen ought to look to*
> *it,—Mr. Saxby knows what he is about,—ED.*

Saxby claimed the lunar theory as his own, but in this, as in other matters, from time to time he was prone to exaggeration. The idea was neither new nor his own. In the eighteenth century, an Italian astronomer, Giuseppe Toaldo, had attributed changes in barometric pressure to the moon's position in relation to the earth. By the nineteenth century the idea was already popular in France. Thomas Bugeaud, a Marshal of France, had acquired an ancient Spanish manuscript whose author claimed that, based on fifty years of observations, he could now foretell the weather with unerring accuracy. And in an 1833 article, "Does the Moon Exercise Any Appreciable Influence on Our Atmosphere?" François Arago argued that the answer was yes, although he later changed his mind. And in England, Richard Morrison (Zadkiel) claimed to have preceded Saxby. "As long ago as 1837," he complained in a letter to the London *Standard*, "I published *The Meteorological Almanac*, in which the theory of changes in the electrical condition of the atmosphere was asserted at the period of the Moon's passing the equator and the tropics."

What's more, a decade before Saxby started putting his ideas on weather to paper, the *Nautical Magazine* itself had even published articles on the subject. In 1850, Royal Navy Commander L. G. Heath wrote, "agriculturalists and sailors have constantly remarked that a change in the phase of the moon is invariably accompanied by a change in the weather," and remarked that at that time in the navy opinion was divided over whether or not there was a connection between changes of the moon and of the weather.

At the *Nautical,* the editors' initial enthusiasm for Saxby's theory began to wear thin, and by the end of the year they were expressing reservations on his method, citing a number of ships that had been lost in an autumn gale that Saxby had failed to foretell in his July prophecies. Nevertheless, they did continue to publish them.

Saxby was passionate about his warnings and sincerely believed in what he was doing, but he felt his was a voice crying in the wilderness. In an 1862 article titled "The Coming Winter and the Weather," he delivered an earnest appeal for people to take his weather warnings seriously. He anticipated a severe winter in which the National Lifeboat Institution and the Shipwrecked Mariners' Benevolent Society, which had run out of money following unusually heavy gales in October, would be particularly needed. And he noted that if more attention had been paid to his "humble suggestions" the prospects for the coming winter would look "more cheering."

"As a member of the British public," he continued, "deeply sympathizing through a whole life with those exposed to the dangers of the sea, (in which, in my own person, it was my lot to have large experience in my younger days,) I cannot...where human misery seems increasing to a fearful extent...lose sight of other threatenings as to the coming winter, which reach us painfully." He was particularly concerned that the public's concern had been diverted from the plight of shipwrecked mariners by the cotton famine in Lancashire, where the American Civil War had interrupted the supply of cotton to the Lancashire mills and had left thousands of cotton weavers destitute—although of course he sympathized with their plight too.

He also changed his mind over the value of FitzRoy's storm warnings, now suggesting that seafarers were disappointed in the Meteorological Department's storm signals. "Why do the seamen on the [East coast] put to sea when the drum or the hoisted cone warns of a gale?" he asked. The answer, he claimed, lay in the number of storm warnings that had not been fulfilled. If mariners thought the department was crying wolf, they were bound to start ignoring the warnings and return to trusting their own judgment of the weather.

Saxby's system had a number of failings from a scientific point of view. What he had observed was merely a correlation, the apparent coincidence of stormy weather with certain periods of the lunar cycle. This of course does not mean that one is responsible for the other. There was no statistical evidence to support it—the science of statistics was in its infancy. It was backed up only by a number of anecdotal reports, and these Saxby seized eagerly and extolled at length, while seemingly ignoring his own failed warnings. Unwittingly or not, he was cherry-picking the results that fitted his theory. For a scientific man, it was a most unscientific enterprise, and his predictions, in reality, were no more reliable than those of the astrologers and the almanacs. Saxby was, quite simply, wrong.

The other great fault with Saxby's theory was that he never specified where his predicted storms would take place. The whole world was his canvas, and his predicted storms might be spawned anywhere on earth, although sometimes he did narrow their limits to either the northern or the southern hemisphere, if the moon was at its maximum declination in one or the other. Saxby's followers in Britain assumed his predictions were intended for them; followers as far away as New Zealand assumed the same. Perhaps he did not know, as we know today, that at any given moment, on any given day, in any given year, hundreds of storms are raging somewhere in the world, and at least one of them is of major proportions. It was of course unreasonable to expect that major weather disturbances would occupy vast areas of the earth at once, but Saxby's followers never seemed to worry about such a minor detail!

And Saxby had built up quite a following. His weather predictions were becoming well known in England. They had also captured the attention of others around the world who had a personal stake in the weather—a significant achievement, given the slow and unreliable means of communication available at the time. Most of his disciples were mariners who were convinced he was right, and who eagerly awaited his predictions. There were also some scientists who, while not prepared to publicly proclaim that the lunar theory had any merits, did not entirely dismiss it, either. Ironically, in view of what was soon to make the name Saxby go down in history, he did *not* think that tropical cyclones—hurricanes—were influenced by the moon.

***

The same month that Robert FitzRoy issued his first storm warning, February 1861, Saxby paid him a visit at the premises of the new Meteorological Office at No. 2, Parliament Street, in London. Saxby was excited about his lunar theory and eager to elicit some authoritative support. He intended to explain the theory to Fitzroy and to persuade him of its value. Fitzroy was a busy man, but he made time for his visitor. He received Saxby cordially enough, but showed little interest in his ideas. He was well aware of the unorthodox theory and had already dismissed it as unscientific, so he did not give Saxby an opportunity to present his evidence. Saxby left the meeting disappointed but not entirely empty-handed. "While [FitzRoy] refused to hear a description of my experiences in the supposed new theory," he wrote, "under the very natural plea that with all the appliances of the country at his command, he could not possibly need any help or information from others, [he] lent me instruments to assist in my further investigations."

Fitzroy was not the only establishment figure to politely rebuff Saxby and his lunar theory. Saxby had earlier written to the astronomer royal, George Airy, who advised him to discontinue his investigations. "I had the misfortune to differ from all the greatest men of Europe in my opinions upon lunar influences on weather," he wrote. "In my idea a man who like myself wrote from *expressed* conviction deserved sympathy and courtesy; and indeed 'real philosophers' thought so too, for with great kindness I was advised by one whom all the world of science reveres [Airy], not to follow up what he considered to be a fruitless pursuit, as the influence of the moon on our atmosphere had been so often tested that various theories connected therewith had broken down when fairly investigated. I shall ever feel grateful to him for his honest reply; others maintained silence; while some, and among them one who is acknowledged to be one of the lights of meteorological science, urged and encouraged perseverance, saying, *that theory offered no obstacles to my views.* But from among all these truly great men not one sarcastic word reached me, notwithstanding my apparent presumption."

Saxby, like FitzRoy, was an extremely busy man at this time. In addition to his teaching duties aboard *Devonshire*, he was at work writing textbooks for his students. His first, on nautical astronomy, was followed a year later by another on marine steam engines. He was also corresponding with his many followers around the world, and he was promoting his lunar theory whenever he had the opportunity. He was not a man to let the opinions of anyone—even eminent scientists or weather professionals—shake his faith.

✺

With all the data he was receiving from his telegraph operators around the British coast, FitzRoy knew that he was now in a position to go a step further than storm warnings. And so, on August 1, 1861, with the allocation of additional staff to his office, he was finally able to issue the first of his "forecasts" to the *Times* newspaper. It was a momentous occasion for meteorology—the first weather forecast published anywhere in the world—but passed entirely without fanfare. Indeed, anyone reading the Meteorological Report in the *Times*—which had now been publishing these reports for several years, giving the previous day's weather conditions from around the UK and a few locations on continental Europe—might easily have missed the brief paragraph tacked below it, which ran to only twenty-four words:

> *General weather probable during the next two days in the—*
>
> *North—Moderate westerly wind; fine.*
>
> *West—Moderate south-westerly; fine.*
>
> *South—Fresh westerly; fine.*

That day, and every other working day for the rest of his life, FitzRoy and his staff started preparing their daily forecasts as soon as they received the day's first weather telegrams at Parliament Street at 10:00 a.m. They processed all the data, and within an hour they had prepared a forecast that was then sent to the *Times*, other newspapers, and other interested parties.

Issuing forecasts was a bold, almost brazen move on FitzRoy's part, and it brought him nothing but trouble. It didn't take long for people to start carping over their accuracy. Even the *Times* was diffident. It continued to publish them, but distanced itself from any responsibility for their content.

For the public, weather forecasting has been a blood sport from its very beginning. When the weathermen get it right, no one spares them a second thought; but when they get it wrong they can expect nothing but abuse or ridicule. And from the outset FitzRoy's forecasts were fettered by the British public's perception that they offered nothing better than the predictions supplied by the ubiquitous almanacs. FitzRoy was careful to distinguish his work from the *predictions* of the astrologers and other charlatans. He had coined the term *forecasting* precisely to make it clear what he was doing and what he was not doing. In speaking of his forecasts he said, "prophecies or predictions they are not; the term forecast is strictly applicable to such an opinion as is the result of scientific combination and calculation." He knew his system was far from perfect, because there were still great gaps in the current knowledge about weather, but it was based on the best knowledge available.

Occasional complaints from the public about the accuracy of his forecasts were only to be expected—the forecasts were necessarily broad in scope and could not take into account local variations and effects—but as FitzRoy struggled to gain acceptance for them, he also found himself up against vocal sections of the scientific establishment, itself struggling to gain acceptance and respectability for the fledgling science of meteorology, and which attacked him for being unscientific. Even Matthew Maury, with whom he had forged a friendship, thought that he should not be issuing forecasts. But it's also true that FitzRoy did have the support of other sections of the scientific community, Astronomer Royal George Airy among them. And he received many letters of appreciation. He was even called upon to give weather forecasts for Queen Victoria when she crossed by boat to the Isle of Wight and for other members of the royal family when they crossed to France.

Some of the fiercest barbs aimed at FitzRoy came from those with their own vested interests. He was attacked by men who saw their hopes of making money from commercial forecasting nipped in the bud by his free forecasts issued in the newspapers, and from a number of callous shipowners who, with little concern for the safety of those manning their ships, did not like to see their vessels lying idle in port while they waited out storms forecast by FitzRoy's office. His most ardent critic, the MP for Truro, Augustus Smith, launched a furious attack on FitzRoy in the House of Commons, and in the newspapers. But FitzRoy was certain that Smith had his own interests at heart: a loss of revenue in the form of harbour dues. Since the storm warnings had come into effect, fewer vessels were finding themselves in distress and putting into port in the Isles of Scilly—over which Smith held a lease.

In an effort to explain more fully the nature of his work, in 1863 FitzRoy felt compelled to put his thoughts on paper. The result was yet another volume in the growing literature on meteorology: *The Weather Book: A Manual of Practical Meteorology*. FitzRoy devoted quite a lot of space to dismissing the ideas of lunarists like Saxby. Much of the *Weather Book*'s content still holds good today, but FitzRoy incautiously let slip a number of ambiguous statements about the lunar theory, and these were to make rich fodder for Stephen Saxby.

≈≈≈

Undeterred by the rebuttals of FitzRoy and Airy, in 1862 Saxby set out to explain his lunar theory in his third book, *Foretelling Weather: Being a Description of a Newly Discovered Lunar Weather System*. He followed it up a couple of years later with a second edition, largely rewritten, and also retitled *Saxby's Weather System: or Lunar Influence on Weather.*

It seems Saxby felt that the scientific establishment had closed its collective mind to the possibility of the moon's influence on weather. Much of his book focuses on circumstantial, anecdotal, and other "evidence" that he insists supports his ideas and the accuracy of his predictions, quoting piles of letters he had received from "the most experienced of observers, naval officers and merchants, navigators, agriculturalists, [and] clergymen in numbers" from around the world. But there were no endorsements from those he most wanted—his peers in the world of science.

One of Saxby's predictions, for example, made in October 1861, warned of a violent cyclone over Britain on November 14, passing from the southward and westward. This particular storm, Saxby believed, would be a hurricane that he had predicted months ahead as crossing the Atlantic about that time. He had extrapolated from the time this particular weather disturbance would form, according to his prediction, using its expected speed of travel to determine the time it would arrive in Britain. And on November 13, 1861, Saxby happened to be visiting the offices of Lloyd's in London when one of the underwriters noticed that the barometer was falling rapidly, which aroused quite a bit of excitement due to his warning. Sure enough, a powerful storm swept over London and England's east coast on the fourteenth, and the Board of Trade (that is, FitzRoy) had failed to forecast it. Despite Saxby's many claims, there were a number of occasions when his predictions of serious weather disturbances were not fulfilled, as the *Nautical Magazine* had already pointed out. He predicted, for example, a serious weather disturbance for November 21–22, 1862. But the daily reports in the *Times* give no indication of stormy weather anywhere around Britain for this period.

Saxby's book also included a rather lengthy critique of FitzRoy's *Weather Book*, pointing out its apparent contradictions,

Stephen Martin Saxby

and that FitzRoy seemed to vacillate over the lunar theory. Initially FitzRoy dismisses the lunar theory out of hand, but later in the book he does make a number of statements that Saxby seized upon and interpreted to show that FitzRoy had not entirely dismissed the concept. Saxby mistakenly believed his theory had made an impact on the Admiral when they had met, and had in fact "startled" him.

"Writing at page 4 he was *not* a Lunarist; while writing at page

244 he *was one!*" Saxby remarks with obvious glee. "In other words, the truth had dawned upon the gallant Admiral in the interim, and with the frankness of a sailor he confessed it—*but in his enthusiasm forgot to state to whom he was indebted for the said truth.*" In practice it mattered little what Saxby thought. Even if FitzRoy did entertain doubts about the moon's influence on weather, he certainly never gave the moon the slightest consideration when he was compiling his forecasts.

Others, apart from Saxby, also charged Fitzroy with being a lunarist. FitzRoy denied this vehemently, but it is easy to see from some of his writing why people might think he remained uncommitted. And Fitzroy was not the only establishment figure to be tarred with the lunarist brush. Even the great Sir William Herschel and his son John were believed—wrongly—to be lunarists. Yet as hard as the scientific establishment—Lord Kelvin among them—tried to discredit the lunar theory, it had become so firmly entrenched in the public imagination that there was no shifting it.

Saxby himself was prone to contradictions, too. He claimed to be keeping an open mind on his theory. "If the circumstances to which I refer can be explained as unconnected with lunar influences," he wrote in the *Nautical Magazine*, "I shall not on conviction feel at all ashamed to confess in your pages my misconception. They are at least extraordinary coincidences." But despite all the evidence to the contrary, he never once conceded that he might be mistaken. He deserves credit, at least, for persistence. It takes courage to publicly express a scientific opinion that goes against the grain, and anyone who does so can expect their peers to turn on them like a wolf pack, tearing their theories to shreds and heaping scorn on their ideas.

Whatever FitzRoy and the scientific community thought of Saxby's theory, an uncritical British public was eager for weather predictions from whatever source. People were using his "weather lists" in Petersburg, Lisbon, the Cape of Good Hope, Australia, the Coast of Africa, and the West Indies. And, in comparison to the predictions made by the many almanacs that were so popular at the time, Saxby's had a certain air of respectability. They were based on what at least purported to be scientific principles.

But *Saxby's Weather System* came in for its fair share of criticism, too. The *Journal of Agriculture*, for instance, wrote: "That philosophers do not assent to this popular opinion is, no doubt, true, not because they deem it antecedently improbable, but because it does not appear to rest on satisfactory evidence. Mr. Saxby, however, writes as if scientific men were in this matter under the influence of 'very strong prejudices, which still hold their judgment in bondage.'" Saxby's response was to fire off a volley of letters to editors of newspapers as far flung as the *Daily Southern Cross* in New Zealand.

The scientist Sir John Herschel completely dismissed the lunar theory: "…the 'weather prophet' who ventures his predictions *on the great scale* [my emphasis] is altogether to be distrusted—a lucky hit may be made: nay, some rude approach to the perception of a 'cycle of seasons' may possibly be attainable…The moon is often appealed to as a great indicator of the weather, and especially its changes as taken in conjunction with some existing state of wind or sky. As an attracting body, causing an 'aerial tide,' it has of course *an* [Herschel's emphasis] effect, but one utterly insignificant as a meteorological cause; and the only effect distinctly connected with its position with regard to the sun, which can be reckoned upon with any degree of certainty, is its tendency to clear the sky of cloud…"

Saxby took exception to the term "lucky hits," but it was a criticism that would forever dog him.

***

By the end of 1864, it appeared that victory for the Union army in the Civil War was imminent, and Joseph Henry started making plans to revive the weather network he had worked so hard to establish at the Smithsonian Institution in Washington before the war began. He started negotiations with the North American Telegraph Association for the use of its telegraph lines, but the hoped-for revival was not to be. Three months before Lee finally surrendered to Grant at Appomattox Courthouse, on January 24, 1865, a devastating fire ripped through the upper levels of the Smithsonian building. Apart from destroying irreplaceable book collections, priceless paintings of native Americans, the institute's archives, and much else besides, the fire gutted Henry's office, destroying all his correspondence and most of his meteorological records.

The huge cost of rebuilding and fireproofing the institution—$125,000—left nothing for the meteorological program. Henry turned to the federal government for funding for a new national weather service, but his appeal fell on deaf ears, and, despite repeated efforts, he was unable to restart his programs. It would be five years until a network was again in operation.

Meanwhile, in Britain, FitzRoy's weather forecasts were also about to suffer a terminal blow of quite a different kind.

∼∼∼

The month of April 1865 was unusually dry and warm in London, with scarcely a drop of rain. But this exceptional spell of balmy weather did nothing to raise the spirits of Robert FitzRoy. For months his wife, Maria, had noticed with growing alarm that his health and the balance of his mind were deteriorating—so much so that the family had moved away from central London to the well-to-do suburb of Upper Norwood, in an effort to improve his condition.

At about a quarter to eight on the morning of Sunday April 30, 1865—a little later than usual—FitzRoy awoke at his home. He arose and made his way to his dressing room, kissing his seven-year-old daughter, Laura, as he passed through her room. It was his last deliberate human contact. At first, FitzRoy did not bolt the door of his dressing room, although he had already made up his mind about what he would do next. A little while later he did lock the door, unfolded his razor, and slit his throat. He did not die immediately and was still alive when his family found him, but he died soon after.

It was an act of utter despair, its roots buried in the number of disappointments that had overtaken him throughout his career. Robert FitzRoy had quite simply reached the end of his tether. The constant barrage of criticism from all sides finally proved too much, and brought an illustrious career to a tragic end. It is clear from an obituary in the *Gentleman's Magazine* in June 1865 that he was held in high esteem by his peers. One of his fellow officers wrote:

> *I knew poor dear FitzRoy from his boyhood; a more high-principled officer, a more amiable man, or a person of more useful general attainments never walked a quarter-deck.*

FitzRoy's tragic death left the burgeoning business of weather forecasting in tatters, and had serious repercussions for the Meteorological Department. His critics, who had vilified forecasting as some kind of black art, now eagerly seized the opportunity to discredit his work, and the Board of Trade and the Royal Society established a committee to investigate the work of FitzRoy's department. The outcome was inevitable, given that the Royal Society had always deemed the whole enterprise of forecasting unscientific and shared a widely held belief in the scientific community that responsible meteorologists should limit themselves to gathering more weather data. FitzRoy had recognized the need for a greater understanding of the forces that drive weather as well as anyone, and he constantly accumulated more and more data. But he also knew that he already had a solid-enough science base from which he could produce workable forecasts that would save lives.

The new committee was headed by Francis Galton and included T. H. Farrer, and both men had a bias against FitzRoy. FitzRoy had dismissed as impractical a signalling device Galton had invented and was trying to promote to the navy, and Farrer was an assistant secretary at the Board of Trade whom FitzRoy had slighted for his bureaucratic back-covering. Their report, condemning FitzRoy's results as wildly inaccurate, was a hatchet job. The report recommended an immediate halt to the weather forecasts, and, though the committee suggested the storm warnings should remain in operation, the Royal Society, which took over the responsibilities of the department, ceased issuing both.

Wresting meteorology from the hands of forecasters and into the hands of the scientific elite was a huge mistake that set back the practical application of meteorology—forecasting—for fifteen years. Even though the public had been critical of FitzRoy's efforts, there was outrage among mariners and harbour authorities over the cessation of the storm warnings, and in the scientific press. The Board of Trade was forced by public pressure to reintroduce them. The dismantling of the weather forecasting service was even criticized by the astrologist Zadkiel, although in life FitzRoy and his department had been the target of his criticism.

Although the Royal Society continued its scientific investigation of weather, it didn't make any real headway for decades. When Joseph Henry in America remarked, "There is, perhaps, no branch of science relative to which so many observations have been made and so many records accumulated, and yet from which so few general principles have been deduced," he could just as easily have been describing the situation in England.

FitzRoy was not fully vindicated until well into the twentieth century, when a re-examination of the Galton committee's blistering report was itself shown to be a sham—"smoke and mirrors" as John and Mary Gribbin denounce it in their recent biography of FitzRoy. Galton's findings were distorted, and the report grossly misrepresented the facts. FitzRoy's forecasts were far more accurate than Galton claimed.

FitzRoy's suicide and the decision to halt official government forecasts must have brought a smile to the lips of the almanac proprietors, who were once again in a position to exercise their fraudulent predictions without competition. The field was also now wide open for Stephen Saxby and his lunar predictions. One of the insults that had been hurled at FitzRoy was that he was the "official Zadkiel." It was an insult that would later be hurled at Saxby.

⌒⌒⌒

Stephen Saxby's first two years at Sheerness passed smoothly enough, and his superiors seemed pleased with his performance. He was expected to teach a wide range of sciences to both engineers and the military branch of the navy, and he had done so with aplomb. Captain Halsted of HMS *Cumberland* (the dockyard commander) wrote a glowing report—Saxby had carried out his duties with judgment, firmness, and ability, and with high credit to himself. Apart from a little encroaching deafness, Saxby's future was looking bright, and he was even admitted to membership of the prestigious British Association in 1863, remaining a member for three years.

It's not clear exactly when or why Saxby's relations with the navy began to turn sour, but it may well have had more to do with the Admiralty's cost-cutting obsession than any other reason. By June

1862 Saxby's duties partially ceased, due to "local difficulties." The following January it was announced that his appointment was to be abolished in March, but that order was rescinded, and Saxby was retained until further orders. The following year, March 1864, the navy proposed that in addition to teaching the engineer officers he should also teach engineering students from the dockyard. Saxby declined, saying rather haughtily that it was derogatory to his position to teach "factory boys." He felt he was above that, and had always refused to wear a uniform. The navy backed off, allowed him to teach only the engineer officers, and declared that the dockyard schoolmaster should teach the others.

There were practical problems in holding classes aboard *Devonshire*. It was often difficult, especially in rough weather, to convey Saxby's students to and from the ship for their lectures. In addition, the navy had found that the cost of Saxby's boat crew was equivalent to the cost of providing him with lodgings ashore. And, finally, the engineers who had attended Saxby's lectures found them too deep and abstruse, preferring the instruction available in the dockyard school, which was more useful in preparing them for their exams. And so in November 1865 *Devonshire*'s shipboard establishment was broken up, and Saxby was provided with a classroom in the dockyard and a lodging allowance, on a temporary basis that depended upon convincing the navy that his lectures were of use. Saxby moved ashore to lodgings at number 7, Albion Terrace, Faversham and rather cheekily, under the circumstances, applied for a pay rise, which was—as he no doubt expected—rejected. His request to be formally confirmed in the rank of Principal Instructor of Naval Engineers (rather than Acting Naval Instructor) was also refused. He now began delivering a course of weekly lectures on a wide range of scientific topics to chief engineers, and daily gave private tuition to any officer who sought it. Saxby settled back into a regular teaching routine, for the time being.

By 1867 the *Nautical Magazine* had a complete change of heart over Saxby's lunar weather theory: "In reference to the moon's effect on the weather," ran an editorial, "we do not believe that she has any whatever, excepting through the influence of the tides…That the stream of the tide influences the weather every sailor knows. The waterman along the coasts of this country will occasionally look for 'more dirt' on the flood, as naturally resulting from his own observations…Still, in our opinion, the moon has nothing to do with the weather." Instead the magazine argued that readers should keep an open mind as to another possibility: the influence of shooting stars on the weather!

Then, in the winter of 1868, Stephen Saxby sat down at his desk and penned a prediction that would immortalize his name. In a letter to the London *Standard*, published on Christmas Day, he started off as usual by giving an example of one of his previous, successful predictions. Then he continued:

> *I now beg leave to state, with regard to 1869, that at seven a.m. on October 5, the moon will be at the part of her orbit which is nearest to the earth. Her attraction will, therefore, be at its maximum force. At noon of the same day the moon will be on the earth's equator, a circumstance which never occurs without marked atmospheric disturbance, and at 2 p.m. of the same day lines drawn from the earth's centre would cut the sun and moon in the same arc of right ascension (the Moon's attraction and the Sun's attraction will therefore be acting in the same direction); in other words the moon will be on the earth's equator when in perigee, and nothing more threatening can, I say, occur without miracle…*

With your permission I will, during September next, for the safety of mariners, briefly remind your readers of this warning. In the meantime there will be time for the repair of unsafe sea walls, and for the circulation of this notice, by means of your far reaching voice, throughout the wide world.

The following September, he duly issued his promised reminder: "in October next all three corresponding causes will occur within the space of seven hours—perigee on the 5th at 7 a.m., lunar equinox at noon, and new moon at 2 p.m." He added that the disturbance would be intensified because the earth was in part of its orbit which would bring it close to the sun. As far as Saxby was concerned, it was a worst possible scenario. But he did not specify where bad weather could be expected: "The warnings apply to all parts of the world; effects may be felt more in some places than in others." Despite his good intentions, it was a qualification that made his warning so vague that, for all practical purposes, it was quite useless.

Hardly had Saxby's warning letter been published in London than his difficulties with the navy resurfaced. In February he was asked to provide details of the number of engineers attending his lectures and how many lectures he had delivered in the preceding three months—a report from the Steam Reserve claimed he had given none. His reply was somewhat evasive, stating that there had been scarcely any engineer officers in port to attend his lectures during this period and that much of his time was spent in giving scientific demonstrations rather than lectures. He also expressed his disappointment that the officers at Sheerness were under no orders to attend lectures, as they were at Portsmouth and Devonport, but did so voluntarily. But the cards were stacked against Stephen Saxby. One of his superiors sent a memo to the controller of the navy, Sir Spencer Robinson:

> *Mr. Saxby's duties are altogether anomalous and no similar office exists in any of the other dockyards. From the mass of correspondence which has taken place respecting him it is evident that the arrangement at Sheerness has never worked satisfactorily, or at least not for the last eight years. The educational requirements of naval engineers as a body are not much superior to what they were formally, as evidenced by the reports of the Director of Education, and such instruction as they require can be given by the dockyard school masters. If this can be done satisfactorily at Portsmouth and Devonport it is evident that it can also be done at Sheerness where a smaller number of engineers are borne in the Reserve.*

> *There can I think be no doubt that the necessity of employing Mr. Saxby as a special instructor of naval engineers has ceased to exist and I beg leave to recommend that the office be abolished.*

In May, 1869 Stephen Saxby's position with the navy was duly terminated, and he was discharged without a pension but with a gratuity of half a year's pay. He was now sixty-five years old, and his wife, Ann, seventy-one. Saxby himself, although angry with his treatment at the hands of the navy, decided that further protest would be futile. His youngest son, Gavin Frank Saxby, however, now a fellow and tutor at St. Augustine's College Canterbury did not, and took up the fight on his father's behalf.

He wrote an impassioned letter to A. J. B. Beresford Hope, the Member of Parliament for Cambridge University, arguing that his father had not received the credit or recognition that was due to him, that his position had always in fact been intended as permanent, and that he was entitled to a pension.

It was no good. The Lords of the Admiralty had made up their minds. Saxby's naval career was over.

If you don't like the weather, just wait five minutes. It may be a tiresome Canadian cliché, but it does have an element of truth—perhaps nowhere more so than in Halifax, Nova Scotia, where abrupt swings in the weather are a way of daily life. No one knew this better than thirty-four-year-old Frederick Allison, who in 1869 was an experienced amateur weather observer living in that city.

As a boy, Allison attended King's College in Windsor, Nova Scotia, where he started taking temperature measurements with a cheap thermometer fixed to the side of a building. At the time few people knew what the most suitable instruments were for recording weather (thermometers, for example, could differ from one another by as much as seven degrees, Allison found), where they should be placed, and when measurements should be taken. But it was a start, and, whenever he could, Allison continued to keep a daily record of the weather. Following the death of another amateur observer, a Colonel W. J. Myers, who had been supplying such records to the Nova Scotia Institute of Natural Science, Allison was asked to fill the void, and from 1867 onward he published his weather observations in the institute's transactions. By now he was equipped with accurate weather instruments at his home on the city's fashionable South Park Street, where he lived with his wife Sarah and was making a comfortable living in the life insurance business. He became the city's unofficial weather man, contributing regular monthly weather reports to a local newspaper.

The need for a forecasting system in Nova Scotia was obvious to Allison. In his first paper to the Nova Scotia Institute he wrote: "The benefits of such a system have been so well proved in Great Britain, and the Continent of Europe, that from me no remarks on its utility is [*sic*] necessary. The advantages gained from forecasts, by commerce and agriculture, have been widely acknowledged, even while, with the data at their disposal, observers stand but at the threshold of a science, which time, accumulating facts in its yearly course, must of itself complete." Allison also wrote of a photographer he knew, who was losing a lot of money because he could never tell what kind of weather to expect on any given day. On mornings when the weather looked threatening he did not prepare photographic plates, often to find that the day turned out clear and he was unprepared for customers. Or on days that started out clear, and he expected to do a good business, it often turned to cloud and rain, and his plates were spoiled. "These frequent mishaps could to a great extent be anticipated," noted Allison, "by signals giving the probable coming weather."

This view was shared by a fellow meteorologist in Toronto, George Templeman Kingston, who had entered that branch of science almost by accident. In 1855 the Cambridge-educated Kingston had been appointed head of the Toronto Magnetic and Meteorological Observatory at the University of Toronto, established by the ruling British government sixteen years earlier. When Kingston arrived in Toronto to take up a new position as a professor of natural philosophy, though, he found there had been a mix-up and the position had been awarded to someone else. Instead, he was offered another post, as professor of meteorology and director of the observatory. Kingston had little choice but to accept, and within a few years had started putting together a weather network in Canada. He received much moral support from Joseph Henry in Washington, and others, but it was to be almost twelve years before he could put his ideas into full operation. Kingston became aware of Allison's work on weather when he was looking for recruits to act as observers, whom he hoped would eventually form the core of a national meteorological service. The two men began to exchange letters in 1869, but, for the time being, Allison continued working on his own.

When Saxby's warning letter came to Frederick Allison's attention, it struck a chord with him. Like most of his contemporary weather men he was not a lunarist, but he clearly did not dismiss the idea that the moon can play a role in the weather either. It's even possible that he had been following Saxby's weather predictions in the *Nautical Magazine* and elsewhere. For whatever reason, he thought there was some substance to Saxby's warning, and he too picked up his pen and issued a letter, which was published in the *Halifax Evening Express* on October 1, 1869.

> *My attention has been drawn to a letter of Capt. [sic] Saxby R.N. to the Standard London, in which a remarkable atmospheric disturbance is predicted for the coming 5th of October, as a result of the relative positions of the Earth, the Sun, and the Moon on that day...I have been asked my opinion with regard to these forecasts; and would thus state it publicly, in the hope of doing some good.*
>
> *I believe that a heavy gale will be encountered here on Tuesday next, the 5th Oct., beginning perhaps on Monday night, or possibly deferred as late as Tuesday night, but between these two periods it seems inevitable. At its greatest force the direction of the wind should be South West; having commenced*

at, or near South. Should Monday, the 4th, be a warm day for the season, an additional guarantee of the coming storm will be given. Roughly speaking, the warmer it may be on the 4th, the more violent will be the succeeding storm. Apart from the theory of the moon's attraction, as applied to Meteorology,—which is disbelieved by many—the experience of any careful observer teaches him to look for a storm at next new moon; and the state of the atmosphere, and consequent weather likely, appears to be leading directly not only to this blow next week, but to a succession of gales during next month. Telegrams from points to the South West of us, might give notice of the approach of this storm; and I trust this warning may not be unheeded.

In addition to its repetition of Saxby's warning, what makes this letter remarkable is what Allison implies, without actually stating it. Warm weather, he said, would make the storm more violent, and it would approach from the southwest—two features that suggest the storm he was expecting was a hurricane. Perhaps a recent hurricane was still fresh in his mind. Just weeks earlier, on September 8, New England had been hit by a rather curious hurricane. A storm made landfall in the region of Long Island at 9 p.m. on that day with wind speeds of eighty knots (148 km/h), and the winds were still strengthening. It then tracked its way across Rhode Island, Connecticut, Massachusetts, New Hampshire, and Maine. Its path of destruction was, thankfully, very narrow, and the winds to the left of its path were hardly more than a stiff breeze. The winds on its right, however, were ferocious, reaching one hundred knots (185 km/h), and Boston lay directly in their path. A number of churches suffered serious damage, and the roof of the Coliseum collapsed when the wind tore out the gables that supported it. The storm was still a power-ful beast when it entered Canada and caused havoc with shipping in the Gulf of St. Lawrence, where it sunk two vessels.

∾∾∾

Despite all the setbacks that had forced Joseph Henry to aban-don his weather network in the United States, all was not lost. On February 1, 1868, a twenty-nine-year-old New Yorker called Cleveland Abbe was appointed director of the Cincinnati Observatory. He was extremely short-sighted, and as a result had been turned down for military service in Lincoln's army at the out-break of the Civil War. But his ideas, like Henry's, were far from short-sighted, and he dreamed of extending the observatory's field of activities to include, among other things, storm predic-tion. It was an ambitious scheme, but he obtained the financing he required from the Cincinnati Chamber of Commerce; secured the cooperation of the Western Union Telegraph Company, for-mer Smithsonian volunteers, and the *Associated Press*; and within eighteen months had cobbled together a patchy but workable weather network.

On September 1, 1869, Cleveland Abbe issued the first offi-cial weather forecast in the U.S., although he did not use the term "forecast." Abbe preferred to call his pronouncements weather "probabilities"—a choice that gave rise to an endearing nickname, "Old Probabilities"—and he made them available to newspapers in several cities. Abbe's system was erratic and unre-liable at first, and there were still vast gaps on the American map from which no observers were reporting.

One glaring gap in particular simply could not be filled at the time. Just as in England it was not possible to obtain weather reports of severe storms approaching from the Atlantic, so, also, was Abbe's system unequipped to provide advance warning to

America's eastern seaboard of hurricanes approaching from the West Indies or the Gulf of Mexico.

<center>~~~</center>

In England, Saxby's warning unleashed a wave of letters to the *Times* newspaper. The warning was, to say the least, controversial, and did not go unchallenged. One of its most vociferous critics was a regular weather correspondent for the *Times* called R. H. Allnatt of Eastbourne, who in the absence of FitzRoy's forecasts had been supplying summaries of the previous month's weather.

Then somehow the emphasis of Saxby's weather warning shifted from expectation of a weather disturbance to expectation of something else that he had only implied: an exceptionally high tide. The discussion prompted Astronomer Royal George Airy to set the public mind at rest. "There is not," he wrote, "the smallest ground for alarm in the height of this tide, as depending on the positions of the sun and moon."

It did little good. The controversy simply drew more and more attention to the forthcoming event, now expected by many to be a "tidal wave." And, when the day arrived, the British public flocked to the country's coastline and the banks of its tidal rivers to take in the spectacle.

There is a certain excitement in the anticipation of watching a rising tide that you know is going to overtop seawalls or other defences. Many were hoping to witness an exciting natural disaster, and to enjoy that age-old pastime of revelling in the misfortune of others.

In Britain the morning dawned calm and bright with a little cloud cover, and warm with light, variable winds over most of the country. Dense fog was reported in the northeast. The seas all around the coast were moderate. The only indication of any weather disturbance was a gale far away to the south in the Bay of

Biscay. There was no sign of the bad weather that Saxby had predicted, and, more to the point for the expectant public, it became apparent as the day progressed that there would be no unusually high tide, either.

Crowds of people had left Preston by train for the coastal town of Blackpool to witness the great tide. Many more had arrived on Tuesday evening for the same reason. Well before noon crowds of spectators packed the beach and promenade—but all they saw was a regular high tide. At nearby Lytham also there was nothing extraordinary. By noon excited crowds had also gathered at New Quay, where loads of clay had been prepared to block the entrances to the boarded cellars if necessary, but at the moment the tide turned the water was still three feet below the sill of the quay, and the disappointed crowds dispersed. Large crowds thronged to London's Embankment and to London, Southwark, and Westminster bridges to witness the anticipated tide. At 2:30 p.m. the tide was full, but only about a foot higher than an average spring tide.

The same story was repeated around the coast of the British Isles. At Chatham, Shields, Preston, Falmouth, Plymouth, and Edinburgh, there were reports that people had prepared for an extraordinary high tide by boarding up the lower portions of their homes, removing furniture to upper stories, and removing goods from harbourfront warehouses, but the high tide was no higher than normal. "All apprehension of any tidal catastrophe is now subsiding," observed a reporter in Falmouth, "the weather being moderate and clear."

Stephen Saxby was no doubt feeling a mite glum. The navy had pulled the rug from under him, and now in the eyes of the British public he was a charlatan whose weather predictions were no better than those of the astrologers or the almanacs. Little did he know that, on the other side of the Atlantic Ocean, nature was unleashing an event that, even if it would not restore the public's belief in his ability to predict storms and vindicate his ideas about the moon's influence on weather, would forever link his name with one of the most devastating hurricanes ever to hit the Canadian Maritimes.

PART TWO
The Storm

One would almost believe that a Monsoon of the Indian Ocean, a tornado of the coast of Guinea, and a West India hurricane, had all agreed to meet in unison, and come down upon our Bay of Fundy shores, with a velocity far exceeding 100 miles per hour.
—Letter to the *Saint Croix Courier*,
October 14, 1869

CHAPTER 5
Deluge

The reading of a storm is not so bad as the feeling of it.
—Cotton Mather, 1724

As far as anyone can tell, the crew of the *Village Belle*, out of Rockland, Maine, were the first to suffer the wrath of Saxby's storm. The schooner had left the island of Inagua in the Bahamas a week earlier, bound for New York with a cargo of logwood and coffee. It had been plain sailing until the vessel was about 180 nautical miles (330 km) south-southeast of Cape Hatteras when a storm of unexpected violence, which had been following in its wake, caught up with it.

The captain may have noticed an increasing swell from the south, and some streaky cirrus clouds scudding northwards a day or so ahead of the hurricane, but even if he had spotted these tell-tale signs there was little he could do as the wind increased but reduce sail to a bare minimum—even bare poles—and batten down the hatches. As the hurricane overtook the *Village Belle* it was hammered by massive winds and tossed around like a piece of driftwood by a tremendous cross sea. The first thing to go was

A nineteenth-century sailing vessel battered by a hurricane.

the deck-load of logwood, swept overboard by tumultuous waves breaking over the deck. Then the foremast broke, soon followed by the bowsprit and the mainmast. The vessel was now helpless before the storm, and breaking waves continued to batter it, smashing the bulwarks and staving in parts of the deck. The men found that the hull had sprung a leak, and took to the pumps to save their schooner from foundering.

Fifteen hours later, when the storm had passed, the crew no doubt offered up a prayer for their salvation. They set to work plugging the leaking hull and cobbling together a jury rig from bits and pieces of spars they had managed to salvage, and what little was left of the tattered sails. Two weeks later, the crippled *Village Belle* finally limped into New York harbour.

In 1869, although scientists were beginning to understand *how* hurricanes formed, they had no clear idea of *where* they formed—nor could they have. It was only the tools of twentieth-century science, like weather radar, and satellite imagery, that have since made this possible.

Many of the most violent Atlantic hurricanes follow a well-established pattern. They are spawned far away in the unlikely setting of the Sahara Desert, when a mass of super-dry, super-hot desert air meets a mass of warm, moist sea air over the Gulf of Guinea, near the equator, and a weather disturbance known as a tropical wave is born. It is a small patch of unsettled weather— an area of low pressure, or a trough. Steady easterly winds blow this wave out over the Atlantic, where, if the sea is warm enough, atmospheric conditions are just so, and it is far enough north of the equator that the Coriolis effect will start it spinning, it will develop into something more serious, a tropical depression; and, again, given just the right conditions, the depression may strengthen into a tropical cyclone—a hurricane.

But Saxby's storm was quite different. It was a late-season hurricane (the season usually runs from July to November)—and hurricanes this late in the year are often spawned by weather disturbances much closer to the place where they will eventually vent their fury. There's no telling *exactly* where this one was born, but most likely it was in the western Atlantic, somewhere to the north of the West Indies.

It may have happened something like this.

Whenever two air masses meet there is usually some sort of conflict, and troublesome weather often results. Sometimes a mass of warm air will climb on the tail of a mass of cold air, forming a warm front; at other times a wedge of cold air will slide underneath a mass of warm air, forming a cold front. These fronts are typical of the storms that commonly develop in the latitudes of North America, the ones that deliver most of the gales that occur throughout the year—and such disturbances are known as frontal storms.

Several days before Saxby's storm, a cold front likely formed somewhere over central North America and travelled eastward across the continent. As the front passed overhead, anyone on the ground would have seen low cloud and heavy rain give way to clear air and light cumulus clouds. The air would have felt a lot cooler, too. The front would then have moved across the coast, and over the ocean somewhere between Florida, Bermuda, and the eastern Caribbean, and would have begun drifting south, where it stalled.

There was a marked difference between the wind direction ahead of the cold front—most likely blowing from the south or southwest, and the cold wind behind it—blowing from the north or northwest. The front would eventually have broken up, but for events taking place much higher in the atmosphere. At about the altitude at which a modern jet plane flies when it's on a long distance route, an area of low pressure possibly moved over the top of the front, and its counterclockwise spinning motion started to drag the lower-level winds—already blowing in almost opposite directions ahead and behind the front—along with it. The winds ahead and behind the front began to dance a little jig around each other, until they began to spin in the typical motion-pattern of a cyclone. As if they needed any encouragement, they were far enough north of the equator to benefit from the Coriolis effect. Within a day or two the cold front would have disappeared completely, with the winds now rotating around a central hub or core, filled with warm moist air. Anyone looking down on the disturbance from a satellite would have seen a great circular mass of cumulus clouds being herded toward the centre of the developing storm.

It was now a tropical cyclone, but a few key ingredients were still needed for the rotating system to grow into a full-blown hurricane. First, it needed a source of heat, and this was supplied from the warm tropical ocean water—at least 80°F (26.5°C) at the surface and down to a depth of 160 feet (50 metres) or more.

Second, it needed a warm, moist atmosphere to supply the thunderclouds now developing with a constant source of rising water vapour, the fuel that drives the storm's "heat engine." Remember James Espy's discovery? Hurricanes derive their enormous energy from the latent, or hidden, heat that is released when water vapour condenses as moist air rises, forming cloud or rain, and as this air rises it draws more air in from its surroundings in the form of wind. And in the eye of the hurricane the air was sinking, generating even more heat by compressing the air below it. Third, it needed its winds to blow evenly and steadily, at all levels from the ocean surface to the top of the thunderclouds. If the upper winds are stronger than the lower winds, they can blow the top right off a developing cyclone, and it will collapse.

With all these ingredients in place, the rotating depression, now a hurricane, was ready to do its worst. It started to move northwards, somewhere north of the Bahamas, its winds steadily increasing in strength, steered by a northward-moving jet stream and most likely another area of low pressure in the upper atmosphere to the northwest. Although tropical storms usually veer to the northeast, this one was blocked by the presence of a massive area of high pressure settled over Bermuda. This particularly strong "Bermuda High" barred the path into the mid-Atlantic, and set the storm's course for its appointment with the Maritimes.

As a hurricane develops it becomes, as we have seen, a kind of self-fuelling engine. Strengthening winds whip the sea's surface into a frenzy, tearing the tops off waves so that it becomes impossible to see where the sea begins and the atmosphere ends. When the water droplets in this cloud of blowing spindrift evaporate, they add even more warm moist air to fuel the heat engine. What's more, extra heat is generated as thick clouds develop, and this warmth encourages even more air to rise and stronger winds to develop over the water. This in turn leads to yet more evaporation, and more warm moisture to fuel the storm. All things considered, as heat engines, hurricanes are extremely efficient at converting the heat energy at the sea's surface into mechanical energy in the form of wind.

But this time, there was another meeting of air masses. Another wedge of cold air moved off the continent and slid beneath the left-hand side of the hurricane. As it did so it lifted the air on the left-hand side of the storm upwards, tilting the top of the hurricane to the right—imagine lifting one side of a doughnut that's sitting on a plate. As the wedge drove farther and farther under the hurricane, it forced the air on the left hand side of the storm upwards, and as a result the air on the right hand side began to sink.

The storm now began to take on the character that it would display as it tracked northward toward New England and the Bay of Fundy. When warm, moist air rises—remember Espy again—it cools rapidly and begins to release its moisture as rain, so heavy rain began to fall in the left (west) side of the storm. And the highest winds in a hurricane are generally at an altitude of about 2,000 feet (600 metres), so when they sink, these strong, counterclockwise winds are brought down closer to the surface, putting the storm's strongest winds on the right-hand side of the storm. The storm also started to change shape. It was no longer circular, but shaped more like a comma, or a teardrop.

As the hurricane moved up the eastern seaboard, dropping heavy rain to the left of its track, out at sea, on the right-hand (east) side its winds were enhanced even more by the northward movement of the hurricane—remember Henry Piddington's "dangerous semicircle" from chapter one? There was also rain on that side, and hurricane-force winds were kicking the sea into a wild frenzy.

By the time it approached the offshore waters of the United States, Saxby was probably a Category 2 hurricane, packing winds near 85 knots (160 km/h). Any ship caught in its fury, as the

Village Belle had discovered, was in for a rough time.

Initially the winds offshore were blowing from the southeast, and it was not long before enormous seas began to build from that direction, but as the hurricane passed by the wind direction shifted to the south, and then to the southwest, and with each wind shift it raised a new set of waves from that direction. When two sets of waves approach each other from different directions they create a confused, tumultuous sea that will toss around any vessel caught in their crossfire like a terrier tosses a rat. Sailors dread this type of confused cross sea. The captain may be steering his vessel to ride one wave when it is hit by a wave from another direction. At best the ship is thrown off track. At worst it is in danger of being knocked down and swamped, or of turning turtle.

***

To the west of this growing storm, in Florida, Georgia, and the Carolinas on Saturday, October 2, there was no sign that a hurricane was developing out over the ocean. The skies were cloudy in Key West, and rain had started to fall in Augusta, Georgia, with a wind from the northeast. But heavy rains began to fall in Virginia on Saturday, and soon after that the skies opened in Washington and Baltimore. It quickly became clear that this was no ordinary rainstorm. It was a deluge worthy of Noah, and its effects would be felt for two days as it headed toward New England, all the while dumping more rain than anyone had ever seen.

The residents of Delaware were the first to feel the full force of the rain. So much fell in Wilmington that the lower part of the city was soon under water. Further north, in Washington, the lower storeys of houses near the canals were flooded, bridges were washed away, and the Tiber river overflowed its banks. The entire reserve force of the Seventh Precinct Police was called out

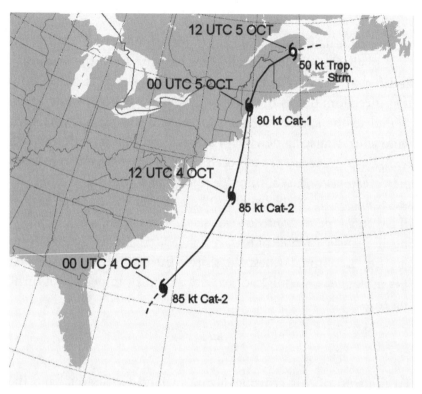

The probable track of Saxby's storm, based on various sources. Courtesy of Chris Fogarty of the Canadian Hurricane Centre.

at midnight to rescue families near Pennsylvania Avenue and First Street, and at times they had to wade in water up to their necks to reach victims. Behind the railroad station the shanties, home to many of the city's black population, were overflowed, but luckily the inhabitants escaped unharmed. Even the dead were not allowed to rest in peace: amid the debris floating down the Tiber were three coffins, unearthed from a burial ground in Georgetown. In Bladensburg the flooding was even worse, reaching the upper storeys of houses. And in Georgetown, where the Potomac was rising at an alarming rate, the Chesapeake and Ohio Canal burst its banks.

By nine on Saturday evening the residents of Baltimore were

beginning to fear a repeat of floods that had submerged parts of their city in July. The Patapsco River was rising fast. The city's firebells sounded a general alarm and merchants began removing goods from the lower parts of their stores. They were only just in time, as the sewers were starting to back up, and the water in Jones Falls Creek rose about two feet in one hour. The lower floors in houses on Gay, Saratoga, and Harrison streets were flooded to a depth of several feet, the foundations of some buildings were washed away, and pavements were torn up.

Some trains were still running, but, as railbeds washed out, most railroads came to a halt, setting the pattern that was to be repeated as the storm tracked northward. In Philadelphia, which seemed to catch the full brunt of the storm, floodwaters from the Schuylkill River carried railcars away, flipping some of them on their roofs. Stores in the city were soon flooded as high as the second storey. And in the river itself, residents were startled to see a great floating raft of houses, boats, cattle pens, large tanks, barrels, furniture, and lumber drifting downstream. On the Lehigh River at Mauch Chunk (now called Jim Thorpe), near Scranton, thirty-seven coal barges were washed over the dam and wrecked, and the iron bridge was swept away, along with several houses. By Monday afternoon the Lehigh River was twenty feet above low-water mark at Bethlehem. Part of town was under water, and the town of Weissport was completely inundated.

By now the floodwaters had also begun to claim human victims. On the Schuylkill River at Manayunk two boys were drowned when their canal boat capsized. Not far away, at Norristown, a woman fell into the river while trying to secure some floating timber, and when a man and woman went to her rescue all three were dragged into the fast-moving current and drowned.

As the rainstorm moved north into New Jersey and New York states, it continued to wreak its havoc. Canal banks were breached, road and rail bridges everywhere were carried off, aqueducts collapsed, there were landslides, and field crops were ruined. Dozens of families living on the outskirts of Jersey City and Hoboken were forced to evacuate their homes.

The Neversink and Delaware rivers reached their highest levels ever known—the Delaware rose twenty feet (six metres), and huge amounts of lumber were washed away, as well as stores filled with the year's harvest of corn and pumpkins. On the Neversink at Cuddebackville the toll house at the suspension bridge was engulfed by the raging waters so suddenly that the occupants barely had time to escape. And at Coleville, about ten miles (sixteen kilometres) from Port Jervis, a woman and a six-year-old girl were drowned when their cart was swept away as the driver tried to cross a swollen stream.

New York City escaped the worst of the rain, although the water in the Hudson River reached a higher level than ever before and the docks and waterfront streets were flooded, forcing their occupants to leave. But conditions were much worse to the west of the city. In Whitehall it rained all day Sunday until noon on Monday. Streams overflowed their banks, the canal burst its banks in places, and a man was drowned, along with his team of horses, as he tried to cross a flooded road.

Hardly a bridge escaped damage in Saratoga—even those that had survived many other recent floods. In Albany, South Broadway became a navigable channel for small boats, and on Elk Street a boy was killed when a sewer collapsed just as his horse and wagon were driving over it. Another boy was drowned at Ballston Spa.

By Sunday evening several dams around Schenectady had burst, flooding nearby streets. The Mohawk River had risen ten feet above its low-water mark, and the Erie Canal and the Delaware and Hudson Canal had both burst their banks, spilling over into the surrounding countryside. Every bridge at Rhinebeck was carried away, and post roads were washed out. The water rose so quickly in Catskill Creek that all the vessels tied up there broke loose, and a great raft of pumpkins and corn washed down the creek and into the Hudson River. At Portchester, a dam burst at an iron works, and the water tore down a machine shop, crushing one man to death and seriously injuring ten others. In Hudson the river reached the highest level ever seen, and at Columbiaville two men were drowned when they tried to save a floating house and were carried over a dam.

Further north, in Rensselaer County, the water rose eighteen feet above low-water mark, on the eastern banks of the Hudson River at Troy. A woman drowned when three houses were carried away at Mechanicsville, and a man was drowned at Fort Ann while trying to save his horse. Two men and a boy were also drowned during a rescue attempt on the river.

As the death toll continued to rise, Rensselaer County became the scene of one of the night's most tragic accidents. A doctor by the name of Fowler and his wife Sarah were among the fifteen passengers heading to the village of Hoosick Falls on the Troy and Boston Railroad when they were involved in not just one, but two train accidents. In the first incident, three railroad employees were seriously injured when the passenger train collided with a freight train near Lansingburg. The passengers were transferred to another train and continued on their way, but at Eagle Bridge the conductor learned that a bridge farther along the track had been undermined and was impassable by anything as heavy as a locomotive. It was still safe to walk across, however, and he knew that his engineer brother had halted his locomotive *George Gould* on the far side of the bridge, unable to proceed across it. All fifteen passengers crossed the bridge safely and transferred to the locomotive, which started to back cautiously toward Hoosick Falls with the baggage master and express messenger sitting on the back of the tender—now the front of the train—clutching flickering lanterns to light its way. The engine ran slowly and without incident until it was only a mile from the town, when the two lookouts suddenly yelled out for engineer to stop. He applied the brakes instantly but it was too late.

Just as the engine ground to a halt, the railbed started to subside, and the train toppled over and plunged down a fifty-foot (fifteen-metre) embankment into the foaming water of the Hoosick River, somersaulting as it descended, finally landing in eight feet of water, into which some of the passengers were thrown. The fireman was trapped in the cab, but managed to free himself. One man from Hoosick was so badly scalded and bruised that his face was unrecognizable. The conductor was thrown clear of the engine and knocked unconscious, and when he came to he found himself lying at the water's edge, pinned to the ground by a rail. Next to him lay the lifeless body of one of the passengers, named Aiken, who had been killed in the fall. Above the deafening roar of the river's torrent he could hear the screams and pitiful cries for help of the injured, and with a great effort he freed himself and went to assist them. But for two of the passengers it was already too late. Doctor Fowler and Sarah had drowned.

Saxby's storm had another surprise in store.

A hurricane will not generally survive for long once it leaves the tropics. It will travel quite far north only if the temperature of the sea surface does not drop below the critical 80°F (26.5°C). As long as the storm follows the same course as the Gulf Stream, all is well; but when the stream veers off to the east in the vicinity of Cape Cod, the northbound storm runs into colder northern water. It will quickly run out of fuel, lose power, and fizzle. It will normally continue to decay as it moves farther and farther north. As it breaks up, the clouds disperse, and all that remains is a gaggle of thundershowers.

Other Atlantic hurricanes start to develop a different character when they leave the tropics, undergoing what hurricane forecasters today call a post-tropical transition. About half the hurricanes that travel as far north as Canada's offshore waters undergo such a transition, evolving into the kind of low-pressure systems that bring our regular gales and thunderstorms.

Saxby's storm would certainly have either decayed or transitioned into a far less powerful beast, but for one thing—a chance encounter with another weather system that gave it a new lease on life. In an encounter like this a dying storm can become explosive with astonishing speed. The kind of tempest that results is no longer strictly a hurricane, nor a typical mid-latitude storm, but a combination or hybrid that can go by any one of an assortment of names: sometimes it's called a weather bomb, or a hybrid reintensification, or an explosive reintensification. Perhaps the name that has lodged most firmly in the public imagination is the one that Sebastian Junger gave to this kind of monster in his bestselling book about the loss of the swordfishing vessel *Andrea Gail* in October 1991—a "perfect storm." One of the more surprising features of such a storm is that it can be even more dangerous than the hurricane that spawned it.

The system that the Saxby storm, now moving rapidly north-northeastward and beginning to run out of steam, barged into on October 4, 1869, was a cold front that was moving from west to east from New York's Hudson Valley to the coast of New England. The front was part of a deep trough of low pressure, and the air ahead of it was in the high sixties Fahrenheit (low twenties Celsius) and heavy with moisture. The air behind it was some 20–30°F (roughly 10–15°C) cooler.

The cold front passed over New York City sometime before noon, and Long Island two hours later. By 3:30 p.m. it was over Boston, and on the coast of Massachusetts it met the rapidly moving but decaying Saxby, now in the vicinity of Martha's Vineyard and outer Cape Cod. When the two systems collided, the cold air behind the front entered the circulating warm air of the hurricane

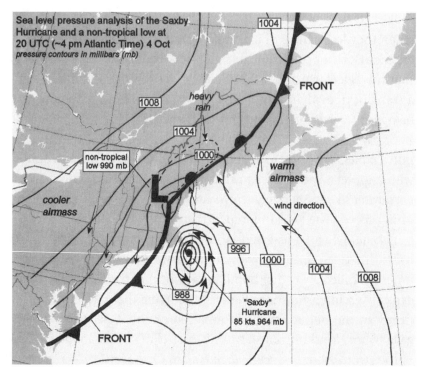

Surface weather map, showing the merging of the hurricane and another low-pressure system in the vicinity of Cape Cod. Courtesy of Chris Fogarty of the Canadian Hurricane Centre, based on Abraham et al., 1998.

and the western part of the hurricane's eye disintegrated. There were also some perturbations in the upper atmosphere, and the net result was that the two circulations joined forces and formed a single, potent, new low-pressure system. Saxby was poised to wreak havoc once again.

Perhaps surprisingly, considering how close they were to the tumultuous events taking place in the atmosphere, and although the barometer at New Bedford and Nantucket fell quite low, neither place experienced so much as a severe gale. The strongest winds remained out to sea where, not far offshore, ships were being battered by the full force of the storm.

Stephen Saxby's warning in the London *Standard* had been taken up by some newspapers in the United States, although not always in time to be of any use. "Capt. Crocker, of Sandwich, and other weather prophets," reported the *Barnstable Patriot* on October 5, "predict a great storm about this time, which shall outblow that of last month, and some of the New York papers advise parish communities to brace up the churches and fasten down the steeples. An extraordinary tide is expected to-day, as the moon will be nearest the earth, and the attraction of both sun and moon will be concentrated on the same point." But by the time readers were browsing the *Patriot*'s warning, it was already too late. The storm had already closed in on Massachusetts. Had a warning been delivered earlier, many Cape Cod fishermen might have thought twice about putting to sea.

But the sea off Cape Cod and Nantucket was, as usual, crowded with ships, almost every one of them quite unaware of the vicious weather system heading their way. The British Bark *Belvidere*, which had left Zanzibar on June 19 with a cargo of spices and ivory, was about 120 nautical miles (220 km) southeast of New York when it encountered the hurricane and spent the next twelve hours battling winds of more than 80 knots (150 km/h) and tumultuous southerly seas. A little while later, and a little further north, off Montauk on Long Island, another British vessel, the bark *Helene*, out of Newcastle, was about to make landfall after a thirty-six-day voyage. It spent fifteen hours struggling to stay afloat as hurricane force winds backed from the southeast to the southwest, during which it lost some sails and split others, had its bulwarks stove in, and suffered other damage.

The *Graham Polly* was about 100 nautical miles (180 km) east of Nantucket when the storm hit. To make matters worse for this ship, it was in the shoal water at the foot of Georges Bank, where a storm can turn a violent sea into a deadly one. Captain Burgess had sailed from New York three days earlier, bound for Glasgow, and when the hurricane hit his ship he experienced a tremendous sea, with winds shifting between SE and SW. When the storm hit it blew away almost the entire suit of sails. At its height, about 8:30 p.m., the ship was boarded by two immense seas. For a while it was completely submerged, and when it finally rose to the surface again, after what must have seemed like an eternity, the seas had carried away everything moveable, including the ship's boats. The bulwarks, fore and aft on both sides, were gone, the stanchions were broken, the forward deck house was stove in, and the contents of the galley and the forecastle, which housed the crew's quarters, were all washed overboard. The *Graham Polly* was in a desperate state, but somehow limped back into New York harbour six days later for repairs.

Not far away on Georges Bank, the storm also caught up with the fishing schooner *Mary Anna* from the Cape Ann region. During the night a terrific sea washed two crewmen overboard, John Welch and Anthony White. In the appalling conditions there was no hope of saving them, and both men were drowned.

Also to the west, but clear of Georges Bank, the ship *Fawn*, out of Cardiff with a cargo of railroad iron, had endured a long, slow passage of fifty days with fine weather until October 4, when, in latitude 41° N and longitude 67° 40′ W, it encountered a violent gale from the south. Captain Nelson hove his vessel to in order to ride out the storm but it too was boarded by a heavy sea, which carried away two quarter boats and the spanker boom, stove in the bulwarks, cabin doors, and windows, and did other damage. Two men were washed overboard, one of whom the crew managed to haul back on board, but the other, a black seaman named Stephen Ephraim, was lost.

Many vessels caught by the storm on the Nantucket Shoals and off Cape Cod reported hurricane force blasts from the southeast, veering south and southwest. A number of ships were driven ashore in Narragansett Bay and Buzzard's Bay where a wind from the south is particularly dangerous. Another vessel capsized at Fall River.

Other ships were even less lucky, and many were driven ashore. The bark *Sarah Elizabeth*, for one, out of Pictou, Nova Scotia, was driven ashore at Gilgo's Inlet, ten miles west of Fire Island Light, and was stuck fast in the breakers. When a crew from the Coast Wrecking Company's salvage vessel *Winants* went to its assistance, their boat capsized in the pounding surf and three crewmen were drowned. Another attempt, this time by men from the Atlantic Submarine Wrecking Company's steamers *Yankee* and *Rescue*, also failed when the surf again proved too heavy for their boats.

It's hard to imagine what the captain of the British ship *Importer* had in mind when he decided to sail from New York, bound for Antwerp, on the night of the fourth, for he was sailing straight into the teeth of the storm. The seas were already mountainous, and as the ship crossed the bar it struck heavily, sprung a leak, and was forced to put back into the city. At Boston, the brig *Wm. Nash*, bound for St. Mare, was also forced to put back into harbour when it lost its sails and bulwarks and suffered other damage. The brigs *Eliza Stevens* and *Flor Del Mar* were both blown ashore, the latter at East Chop. The schooner *Thomas B. Smith* went ashore at Providence, as did the schooner *Juniata Patten* at Watch Hill Reef. There were many other casualties of the storm, including many dismasted ships, and the ocean surface in the wake of the hurricane was like a watery lumber yard, strewn with the deck cargoes and other flotsam swept overboard by the tumultuous seas.

ᴑᴑᴑ

New England was still reeling from the effects of a hurricane that had struck on September 8. Before that, inland parts of the region had been spared the wrath of any tropical storm for forty-four years, and people were hardly prepared for another one so soon. And now, as Saxby's storm moved over Connecticut, a paper-mill dam at New Haven was breached by rising floodwaters. At Hartford another dam was washed away, as was an entire cotton mill and several houses. Manchester probably suffered more than anywhere else in the region. At South Manchester a large reservoir burst, damaging paper mills, a silk mill, and machine shops. Not one bridge was left standing in town and the roads were nearly all impassable.

A young couple and their five children, who lived in an upper-floor tenement of an old factory at Broad Brook, were all drowned when the entire factory was swept away by floodwaters. And as more and more dams and bridges were swept away at Manchester, South Coventry, Stafford, and Birmingham, the waters claimed even more victims.

At Canton, northwest of Hartford, a place known for its heavy rainfall, a weather observer recorded an incredible 12.25 inches (31 cm) of rainfall during the storm. The poor man could hardly believe his eyes and thought he had made a mistake with his measurement. But it was no mistake. The rainfall was breaking all records.

The city of Boston did not escape its share of rain, and here, as the downpour began to ease off, the wind picked up almost to a gale. For the second time within a month, part of the roof of the Coliseum was blown off. The tides in the harbour reached an extraordinary height, stopping just short of the tops of the wharves. At Concord hundreds of cellars were flooded, streets washed out and gullied, and sidewalks caved in. At Holyoke the Connecticut River rose twenty-one feet above low-water mark, and was soon filled with enormous rafts of lumber brought down from the north. Railroads and bridges continued to suffer the worst. In Franklin County alone more than twenty bridges were destroyed. Every mill in the city of Lawrence was flooded, as well as a large part of the city.

When Richard Barryman turned up for his first day of work at the Hoosac Tunnel in western Massachusetts, on Monday October 4, he had no idea it would also be his last. The tunnel, almost five miles long, was being cut through Hoosac Mountain in Massachusetts, to provide a rail link between Boston and upstate New York, and in 1869 it was still several years from completion. It had already taken quite a toll of men's lives. On October 19, 1867, thirteen men had been trapped and killed when a fire destroyed the pumps that drained the workings. The excavation flooded, the men drowned, and it was a year before the tunnel was pumped dry and the bodies recovered. Now, just one year later, it was about to flood again.

Barryman was one of seventy-five miners at work that day, drilling and blasting their way through the mountain, spread out along about a mile of the tunnel. The floodwaters had still not abated throughout much of the region after the weekend's storm, and the rivers and canals were still way above their normal levels. About ten that morning a swollen stream to the north of the tunnel burst through an embankment, and floodwaters gushed into the cutting that led to the tunnel's west end. A side tunnel that was supposed to drain water from the cutting was closed, for some reason, and the water rapidly rose. Torrents of water forced the men in the main tunnel to drop their tools and run for their lives. Their only hope of escape was through a vertical ventilation shaft drilled down from the top of the mountain, and the men fled in that direction, now in pitch darkness, pursued by the flood. All but one made it to safety. A twelve-year-old boy, John Ryan, nearly drowned but was saved by one of the contractor's sons, John Hocking, who himself was swept away by the current three times but managed to finally grab the boy and pull him to safety. The tunnel was rapidly filling with water by now,

Workers in the Hoosac Tunnel. Many narrowly escaped when the tunnel flooded.

and Richard Barryman, unfamiliar with the tunnel and perhaps slower to find his way to the shaft, was seen clinging to a piece of floating timber. It was no good, and the last that was heard of him as he was swept away by the current were the prayers he was uttering as he sank, never to resurface.

There was no sign of a let-up in the deluge when the storm reached New Hampshire. Four inches of rain fell in three hours at Concord on Sunday alone, and it was still pouring at two in the afternoon on Monday. The Merrimack River was rising rapidly. In all, the storm dumped eight inches of rain on the town, submerging large parts of it beneath a foot or two of water.

As bad as conditions were, if anyone living on the coast needed a demonstration that the situation was worse at sea, it was delivered at Newcastle. Two residents were walking on the beach at ten on Monday evening, when they caught sight of a massive wave rolling ashore. They fled toward higher ground, but one of them stumbled and fell. He was able to cling to the rocks where he had fallen as the wave washed over him and rolled up the shore, reaching a point one of the men estimated at 125 feet (38 metres) beyond the normal high water mark. Their story may well have become exaggerated in the telling, but they claimed this "tidal wave" was 18 feet (5.5 metres) high.

∽∽∽

Early on Monday morning the Swift River in Maine appeared quite normal, despite the heavy rains. As its name implies, though, the river is fast-flowing, and when torrents of water run down from the Appalachian Mountains its level can rise rapidly and without warning. So, after all the rain, it was no surprise when the river suddenly started to swell about 8 a.m. Within half an hour it had risen eight feet. The freshet rapidly overwhelmed houses, barns, cattle, and crops. As one man walked along the river to watch the incredible sight, he removed his coat and hung it in an apple tree. When he returned half an hour later to retrieve it, the tree was in the middle of a raging torrent, the swirling water within inches of the branch where his coat hung.

By the afternoon whole trees, torn from their roots, were being swept down the swollen river, demolishing everything in their path. Barns, houses, horses, cattle, hay, and debris of all sorts were carried away on the flood, even a whole farmhouse that had been lifted off its foundation intact. It presented an eerie sight to one witness, who noticed that its contents were still in place—he could clearly see the chairs, beds, kitchen stove, flour barrel, and tin plates through the open front door from which the occupants had only recently escaped. A moment later the house struck a massive tree, and was dashed to splinters.

The rain continued all day Monday, and by 8 p.m. the stream had risen an astonishing thirty-six feet. Further downriver, two lumbermen went to the rescue of an elderly couple whose house was rapidly becoming inundated. The woman, almost eighty years old, was not agile enough to leave the house, and by now the floodwater was up to her arms. So one of the lumbermen pushed her up into the branches of a tree and told her to stay put. He then swam through the fierce current back to dry land. The following morning the men returned to the scene to find the old woman still perched precariously in the tree where they had left her. She had climbed as high as she could, yet even here the waves had splashed her face. After a night spent shivering and thinking she was about to die, she was soaked to the skin but had not suffered a scratch.

∾∾∾

In the eastern States, rain by the bucketful had been the hallmark of the storm until now. Only those unlucky to find themselves at sea with nowhere to run to had felt the fury of hurricane-force winds. But as the storm churned on northward, the winds were coming ever closer to the Maine and Canadian coasts. Those winds were about to come ashore.

CHAPTER 6

Landfall

When the rain comes before the wind,
Look out, and well your topsails mind…
　　　　　　—Traditional seaman's weather saying

Anyone in the city of Saint John, New Brunswick, taking stock of the weather on the morning of Monday, October 4, would have seen nothing to suggest that a powerful storm was approaching. The day dawned foggy, but the fog had cleared by seven and the rest of the morning was cloudy. The only unusual features were the high temperature—by mid-morning it had risen to 70°F (21°C), where it remained until late afternoon—and a heaviness in the air due to the high humidity. But then again, it had been hot for several days now, more like July than October.

Those who had read Saturday's *Daily Telegraph and Morning Journal,* which had reprinted Saxby's warning letter to the London *Standard* under the headline "A High Tide Coming—Look Out for Tuesday Next," were probably scratching their heads. This Indian summer looked like it would never end. Only those familiar with tropical cyclones, or who had also chanced to hear of

Frederick Allison's warning about warm weather, might have suspected that things were about to take a turn for the worse.

When the paddlewheeler *New York*, belonging to the International Steamship Company, slipped its moorings at Saint John and steamed into the Bay of Fundy, bound for Boston, the weather was still fair, although a stiff breeze had picked up from the southeast. Everyone on board was looking forward to an uneventful run for the vessel, one of three steamers—the others were the *New England* and the *New Brunswick*—that plied between Saint John, Portland, and Boston, leaving Saint John every Monday, Wednesday, and Friday morning. One of the other steamers was to leave Boston on the same morning, bound for Saint John, normally arriving on the evening of the following day.

The trip as far as the first port of call, Eastport, Maine, was routine, and the *New York* docked there about 12:30. Passengers might have noticed that the wind had increased steadily throughout the short passage, although this was nothing unusual. But on its arrival the ship took delivery of a telegram from the company's agent in Boston, warning that the wind there was blowing a gale. At Eastport the vessel took on a large quantity of freight, but about three o'clock in the afternoon, as Captain E. B. Winchester was preparing to leave, he received another telegram from Boston advising him that the gale had grown furious in its intensity. He was instructed not to proceed to Boston, but to leave Eastport and find a safer harbour.

Captain Winchester decided that nearby Rumery's Bay, surrounded by tall hills, was his best bet. He wasted no time in raising steam. It was only a short run and the *New York* arrived about 4 p.m., but the wind had already picked up mightily and had reached hurricane force. A number of coasters loaded with lumber had already sought shelter in the bay, but there was still plenty of room for the steamer. The captain called for the main anchor to be lowered, with 45 fathoms (80 metres) of cable, followed by a second anchor with 35 fathoms (65 metres). The anchors seemed to take a good hold and the vessel seemed secure, lying in the lee of the high land. Captain, crew, and passengers felt safe in the knowledge that they would be able to ride out the storm safely in this snug anchorage.

The wind blew harder and still harder. The rain was pouring down in torrents and the *New York* began to rattle and shake in the violent gusts. "I never felt such wind," one passenger later recalled. "It seemed to come in squalls, which you would think would carry away everything." Darkness fell, and about 7 or 7:30 p.m. the strong steel fastenings that secured the hurricane deck (a lightly built deck over the steamer's main saloon) started to work loose. The deck began to separate down one side of the vessel, and the roof and sides of the main deck, said a passenger, were "swaying and bending like cardboard." It seemed only a matter of time before those would be torn away, exposing the hundred or so passengers to the full force of the hurricane.

At great personal risk Captain Winchester and one of his officers ventured out on the hurricane deck. Clinging to whatever handholds they could find to avoid being swept away by the powerful wind, they managed to fasten lines and tackles around the wildly flapping structure.

Meanwhile, in the staterooms down below, the ship's officers were tossing life preservers to the terrified passengers, many of whom had by now turned pale and were screaming in fright. Some were kneeling in prayer. The passengers were so panicked that a number of men were seen clutching more than one life preserver while women and children were still without, although other passengers soon shamed them into sharing.

The same gust of wind that had loosened the hurricane deck also caused the vessel to lurch backwards so violently that the anchors lost their hold, and they were both torn out of the seabed. For a while the steamer dragged backwards at a terrific rate,

The paddlewheeler New York *in 1869.*

but suddenly the anchors seemed to catch again as if on a ledge, and checked the vessel's hell-bent journey through the churning water. But the anchor chains were not able to withstand the terrific shock, and they snapped—first one, then the other. Without anchors the *New York* was now in extreme peril. Even with the engine running at full power, Captain Winchester was unable to keep the ship's head pointing into the wind, the only way to preserve any kind of safety against the violence of the tempest.

The bow was blown off the wind, and as the vessel began to move sideways it almost immediately struck some rocks. It appeared to be held by the stern, and started taking on water around the rudder post. But then, as suddenly as it had struck, the *New York* drifted loose again, stopped taking on water, and glided into a bay or creek known as Lubec Neck. Down below, stokers were piling wood as fast as possible into the boilers to raise every ounce of steam they could. Even the deck cargo lumber was tossed in. The intense darkness was relieved by flames leaping from the top of both smokestacks, giving the scene an almost surreal air.

The engine, however, proved powerless against the hurricane-force winds and the *New York* continued its out-of-control passage, until eventually it grounded again, fortunately this time on a muddy bottom. The passengers and crew could just make out, by the light of the flames from the smokestacks, a handful of small fish houses on the shore. The engine was still running ahead but it could barely keep the steamer from driving further ashore. Captain Winchester sent out a boat to take a line from the forward hawse pipe ashore, but although the overpowering wind and waves drove them back repeatedly, eventually two crewmen managed to jump overboard in shallow water, struggle ashore, and secure the line. The passengers at last began to feel some relief, as they were now so close to land that if worse came to worst they imagined they could jump overboard and swim to shore. And then the hurricane deck, which had been lashed down with lines and tackles after it had started loose and was still flapping around like a sheet of cardboard, finally gave way with an almighty crash and disappeared into the murky blackness.

The vessel remained stuck fast until about ten o'clock, when at last the wind started to subside. Another line was run ashore to some trees and, with all hands to the windlass, a small jib hoisted, and the engine running hard, the Captain was able to bring the bow around and into the channel. *New York* was again heading for open water.

No sooner had it begun to make way than Captain Winchester discovered that the ship's wheel was now useless. When *New York* had grounded the first time the rudder had broken off, and it was now hanging loose, held only by the steering chains. A crewman was charged with sounding the ship's whistle constantly, in the hope that another vessel would come to the *New York*'s assistance. But no help came. Every other vessel in the bay was struggling for its own survival amidst the tremendous storm.

The situation now seemed even more desperate. Although the worst of the storm seemed to have passed and the wind had abated to some extent, without rudder or anchors the vessel was completely at the mercy of the still-violent wind and sea. But by some means Captain Winchester did manage to steer the *New York*. How he did so is anybody's guess, for the ship's side-wheeler paddles could not be driven separately. Yet, by an extraordinary combination of seamanship and good luck, he was able to guide the crippled vessel all the way back to Eastport, where it docked on the stroke of midnight—at the very same wharf it had left some nine hours earlier.

Other vessels that had taken shelter in Rumery's Bay were less fortunate, and by morning seven of them were on the beach, including *Flora King, Lookout, Frances Ellen, Boston,* and *Red Beach*. Of twenty-six vessels in the lower bay, twelve were dismasted and the *Julia* and the *Martha* were badly damaged.

***

At about the time that the *New York* was struggling to maintain an anchor hold in Rumery's Bay, a Nova Scotian vessel, the brigantine *Annie*, was on passage from Margaretsville for Boston. Its captain was worried. Not only did he have his cargo of cordwood and a crew of five to consider, he was also carrying eleven women and four children as passengers on this trip, and he now found himself overtaken by a storm more intense than any he had encountered before. He knew he was about twenty miles (thirty-two km) southwest of Mount Desert Rock, off the coast of Maine, and on a clear night he would have just been able to make out the steady pulse of the beam from its lighthouse, but in the spume-and rain-filled air that now drove past him in sheets he could barely see beyond the bow of his vessel. His worst fears were realized a little later when the wind took away the *Annie*'s masts. His vessel and its human freight were now at the mercy of the wind and waves. All he could do, if he was a religious man, was pray. In the darkness, and unseen by the captain, all around the *Annie* other vessels were also being dismasted, and no doubt their crews too were praying for deliverance.

ааа

It was shortly after dark when the storm finally barrelled ashore at Eastport on the border between Maine and New Brunswick. Residents of the town who had been keeping up with the news—if news had reached this far before the telegraph lines were brought down—were no doubt expecting the kind of torrential rains that had caused so much flooding in every eastern state from Virginia to New Hampshire. They certainly weren't prepared for what actually hit them.

Wild storms are not uncommon in the Bay of Fundy in October, but seldom are they as savage as this. The winds by now had reached Category 2 on the Saffir-Simpson scale (see Appendix 1), and were blowing at speeds of up to ninety knots (166 km/h). Buildings near the water were the worst hit. Those around the International Steamers wharf were completely demolished. Nearly every wharf and every vessel that was tied to it were wrecked; nearly every house along the wharf was blown down; and nearly every tall chimney in the town was toppled, 108 of them in total. Trees were torn up by their roots. Fourteen vessels were stranded, including a large schooner blown high up the beach. Many were total wrecks. Within the space of an hour or so, the whole harbour was covered in a seething flotsam of damaged vessels, lumber, barrels, masts, sails, and fish.

In one house in the town, two young girls were upstairs in bed asleep when they were startled awake as the roof above them simply disappeared into the night and they were left gazing into a coal-black sky. If they could see anything at all it would have been cloud scudding past so low they might have felt they could reach up and touch it. Another family, fearing for the safety of their house, took refuge behind their barn. They realized that the high-sided building was no safer only when it began to move bodily before the storm, finally collapsing an instant after the father had snatched his children out of harm's way.

At nearby Lubec it was a similar story. Many fishermen lost everything they owned as the storm wreaked havoc among their fish houses. The church lost its steeple, and a great many houses and barns here, at Robbinston, at Perry, and in the surrounding countryside were flattened. Further inland, at Calais, where the wind had started to become extremely violent about 6 p.m., it continued to pick up until about ten, by which time hardly a barn was left standing, several people were seriously injured, and livestock were being killed by falling buildings.

For vessels that had sought shelter in the nearby bays and inlets the peril was even greater. There was one hazard in particular that no captain, however good a seaman, could do anything about. Even if his own anchors held, if vessels upwind dragged theirs, he was a sitting duck. The captain of the *Phoebe Ellen* discovered this to his cost, as his log reveals:

> *At 2 PM came to anchor in West Quoddy Bay; about 24 vessels in the bay at the time—at 5 PM blowing a heavy gale, let go second anchor—at 7 PM fearful squalls with gusts of rain—about 9 PM a schooner said to be the* D. Sawyer *of Jonesport drifting past, struck us several times, damaging our bows before*

they got clear. The brig Whittier *came against our bow, crushing gear and bows of our vessel, and then came round our port side carrying away foremast and gear, with main topgallant mast, and her anchor catching ours did not get clear until 5 PM Wednesday evening. On Tuesday morning counted 18 vessels at anchor in the bay.*

No one was keeping count on the night, but a reporter from Saint John later tallied the number of vessels that had been driven ashore on the short stretch of coast between Machiasport, Maine, and St. Andrews, New Brunswick—some of them wrecked, others salvageable: Machias Port, fourteen; Machias, eleven; West Quoddy Bay, eighteen; Rumery's Bay, five; Broad Cove, nine; Eastport, ten; Deer Island, twenty; Indian Island, one; Clam Cove, five; St. Andrew's Bay, two; Calais, St. Stephen, and on the river, twenty-six. In all, 121 vessels were beached. Other small fishing boats simply disappeared in Passamaquoddy Bay without trace.

"Eastport is reported as nearly demolished," the Western Union telegraph operator at Bangor, Maine, tapped out on his Morse key at the height of the storm. With telegraph lines down in every direction, it's a wonder he had heard any news from Eastport at all.

As the storm crossed into New Brunswick there was no let-up. The first town in its path was St. Andrews, where it continued its demolition work, blowing down entire buildings, lifting others from their foundations, removing roofs and chimneys, shattering windows, and downing telegraph lines. The wind was so powerful that it set two loaded railcars in motion and derailed them. People abandoned their houses for fear of being buried alive under collapsing structures, but it was hardly any safer in the streets—pedestrians could barely stand up outside without fear of being blown off their feet. One man, James Perkins, who ventured outside his house and decided to cross the street, was a little surprised to find himself blown across to the other side before he had so much as taken a step, shaken but unharmed. No doubt he kept his head down after that, for all around him the air was filled with a lethal barrage of bricks, signboards, fencing, and other impromptu missiles.

In the harbour, in addition to local boats, others had put in during the afternoon to shelter from the storm. The harbour master, Captain Balson, warned the masters of all the vessels in the harbour to make sure their anchors were secure; but, even so, many dragged and were driven ashore, including the *Utica*, *Calvin*, *Ellen Frances*, *Julia Clinch*, *Truro*, *Emma Pemberton*, *Harrie*, *Mary Budd*, *Elizabeth Bowlly*, and *Matilda*. A similar list could have been drawn up almost anywhere along the coast that night. Several vessels were also dismasted, and the beach was littered with the remains of small boats broken up by the storm. Only five vessels managed to ride it out in safety.

Despite the appalling destruction, no lives were lost in the town, but in the middle of the night, as the winds subsided, the residents were presented with an eerie spectacle in the bay: a large schooner was seen adrift near Deer Island—it was either dismasted or bottom-up, no one could make it out for sure in the deep gloom. It was the hull of an American vessel from Jonesport, named *Rio*. There was not a living soul aboard.

At St. Stephen and Calais, where the St. Croix River meets the head of Passamaquoddy Bay, houses, shops, mills, barns, churches, and railcar sheds were blown down or had their roofs blown off, and horses and cattle were killed. Trees that had defied storms for half a century were uprooted or broken off a few feet above the ground, and fences and shutters dashed to pieces.

More vessels were driven ashore in the river. The steeple of the Episcopal church at St. Stephen began to rock in the wind so wildly that its large bell began to toll. Residents nearby thought it was a fire alarm. Finally the tower lurched and came crashing to the ground. The bell, as legend has it, rang twice as the tower toppled, as if sounding its own death knell. Other houses of prayer were not spared either. The Catholic chapel at Cork Settlement was demolished, and the roof of the Methodist church was badly damaged.

A few miles east of St. Stephen, a Mr. Shaw was crossing a bridge at Waweig with his horse and wagon during the storm when his horse faltered. Shaw climbed down to check what was wrong and discovered that some of the planking of the bridge deck was missing, torn up by the wind. When he turned round to retrace his path, he found to his dismay that several planks were now missing behind him as well. He hitched his horse to a railing and managed to clamber ashore across one of the bridge's stringers to seek help. When he returned the next day, his horse and wagon were no longer there—nor was the bridge.

That night the entire coastline of Charlotte County became a graveyard for ships and men. At Musquash two fishing vessels were driven ashore, one of them the *Renown*, and two men from each were drowned. A Nova Scotia vessel from Big Salmon River with a cargo of spars capsized off Ragged Point, where it came ashore, its crew of five clinging to the upturned hull, their clothes ripped from their bodies by the sea. They were tossed ashore with barely a garment among them. Three vessels were driven ashore at Lepreaux, one a total wreck, and at Beaver Harbour four men were drowned and three vessels driven ashore. The village of Beaver Harbour was itself almost destroyed by the wind and waves. At L'Etang Harbour another young man was drowned when his vessel was wrecked.

Grand Manan Island, which stands guard over the entrance

to Passamaquoddy Bay and also lay directly in the storm's path, offered no defence against the fury of the storm. About a hundred buildings and nearly every boat on the island were destroyed. The wind here was so strong that it blew anyone unlucky enough to find themselves outside clean off their feet. The keeper of the island's Gannet Rock lighthouse described the situation somewhat wryly: "I think the inhabitants never manifested such a strong desire to remain on Grand Manan since its settlement as they did that night. The paramount thought with each was to get hold of something that would not blow up, pull up, or tear up, and happy was the man, woman, or child that night, that could find their way to an alder swamp where there were no large trees to crush them."

At first people feared the island had been the site of a terrible tragedy, as a report of 150 bodies washed ashore at Flagg Cove began to circulate. When the *Saint John Daily Telegraph and Morning Journal* later dispatched a reporter to check, he heard numerous other accounts of horrendous loss of life, including one from a man who claimed he had personally helped pile up a hundred bodies at Lubec, that he knew of thirty others that were picked up at Eastport, and that more were being washed up by the hour. All of these reports proved false. The 150 bodies turned out to be a hoax, and a report of the destruction of the Gannet Rock Light also proved untrue. Several schooners had come ashore though, and nine bodies were recovered from the island's shore. One of these, the *Echo*, came ashore in pieces, its captain, a man named Swift, and his son and the rest of the crew having all gone to a watery grave.

The other islands guarding the mouth of Passamaquoddy Bay—Deer and Campobello—fared no better. And on Bliss Island, near Blacks Harbour, where the schooner *Rechab* had arrived safely at 2 p.m., it was driven ashore. As the vessel struck, the captain, a man named Mowatt, grabbed the peak halyards and swung ashore, but just as his feet touched the ground the vessel lurched and he swung back aboard again. On a second attempt, as the schooner rolled he swung again, but this time he let go of the rope as his feet touched the rocks. The halyard swung back aboard, and the rest of his crew were then able to swing ashore in the same manner. Three other schooners, the *Rosalee B.*, *Ellen McLeod*, and *Gipsey* were also lost in Bliss Harbour that night, and a fourteen-year-old boy, the son of a local captain, William McLeod, was drowned.

Captain James Meeley of the two-masted schooner *Linnet* was in New River Harbour, about thirty miles northeast of Eastport, when he realized that bad weather was in the offing and anchored under the sparse shelter of a small islet that he knew as George's Island (today it is shown on maps as Mink or Mole Island). The name "harbour" is an exaggeration for this place. An adequate anchorage in fine weather for vessels that arrive to load lumber from the mill at New River, the harbour, which lies almost midway between Passamaquoddy Bay and Saint John, is really just a small bay, and except for the slight amount of shelter provided by the island it is wide open to the south and west. It is certainly no place that a sea captain would wish to ride out a hurricane, or any other violent storm.

Captain Meeley had just completed loading his vessel with a cargo of laths and was making ready to sail for Eastport, but when he realized he was in for a storm he set out his anchors with particular care, battened down the hatches, and hunkered down to ride out the worst of it. He must have known that this was far from an ideal anchorage, but he had little choice. The only safe anchorages were too far away for him to reach now. There were two other people aboard, both passengers, a woman and a young boy.

Meeley noticed only one other vessel in the anchorage: the much larger three-masted barque *Genii*. It was a brand-new vessel, 120 feet (37 metres) long overall, of 380 tons registered, its two forward masts square-rigged and the aft one rigged fore and aft. It was the kind of vessel that was common hereabouts—barques were the workhorses of the transatlantic trade. *Genii* had been launched only three weeks earlier, at the yard of R. Glenn & Co. in the busy shipbuilding town of St. Andrews. It had sailed from St. Andrews in ballast on the previous Friday with a skeleton crew of only the captain—an experienced shipmaster, Charles Bayley of Westport, Brier island—a mate, a second mate, and a steward. It was to take on a cargo of sixty thousand board feet (142 cubic metres) of deals—fir or pine timber sawn into planks—at New River Beach, and was thence bound for Liverpool, England, on its maiden voyage. The cargo had been formed into a raft and was secured in the lee of a breakwater, awaiting loading on to the *Genii* on Monday morning. The loading was to be carried out by a gang of seven stevedores who had come aboard over the weekend and were to remain on board until their work was complete.

On Monday morning the stevedores had set about dumping ballast over the side to make room for the cargo. In view of what was about to happen this was unfortunate, as the *Genii* was now riding high in the water, offering more resistance to the wind and thereby putting more strain on its anchors and chains. It also made the barque lighter and more prone to be blown about from side to side, again adding to the strain on its anchor warps. The stevedores completed offloading the ballast by noon and decided to row ashore and take some liquid refreshment at a drinking establishment whose proprietor was a fellow called Mick Haggerty. As the men rowed ashore the sea was dead calm, but by late afternoon, when they returned to the *Genii*, a breeze had picked up from the southeast and the water in the harbour was becoming decidedly choppy. The pilot, James Clark, for

New River Beach, the scene of the Genii *tragedy. The island offers little or no protection from a powerful storm.*

some reason stayed ashore. It was common for the stevedores to remain ashore at night too, and lodge with a local family when they were working at New River, but on this occasion these men had decided to spend Monday night aboard the barque.

Genii was lying somewhat ahead of Captain Meeley's *Linnet*, and on its weather bow, about halfway between the island and the shore. Captain Meeley noted that *Genii* was secured by two anchors, one well ahead of the ship and the other half way between the first one and the ship. The *Genii* had let out plenty of anchor chain, which would be needed now that the wind was picking up. When it did so, Meeley set another anchor from *Linnet*—a kedge anchor—and hauled his ship further into the lee of the bar off Mink Island.

By the time night fell over New River Harbour the wind had risen to a howling gale. Although Captain Meeley could no longer make out the shape of the *Genii* in the pitch darkness, he could see its lights, and they showed him that it was dragging its anchors, slowly at first, and then faster, but suddenly coming to a halt again as the anchors took a hold once more.

At this point the gale had become quite furious, *Linnet's* anchors also started to drag, and Captain Meeley became preoccupied with taking care of his own vessel. For the next half hour he let out more cable, and eventually the anchors bit into the seabed once more. Now that he felt safe once again from the violence of the storm, at least for the time being, he turned his gaze once more toward the *Genii*. But, as hard as he peered into the murk, he could see no lights where the lights of the *Genii* ought to have been.

Genii's anchors can have held firm for a only few minutes after they took hold, and once again the furious wind caught its hull, masts, and rigging, and the anchors broke loose. This time the vessel was beyond hope. It was driven helpless before the tempest until finally it struck, broadside on, the westward pinnacle of New River Ledge, a black outcrop of jagged rocks, normally just visible at high water, but now covered in foaming breakers. All three masts broke like matchsticks under the shock of the impact. As the huge waves continued to break against its side the *Genii* rolled over onto the ragged rocks and turned turtle. The men, who were on deck, were hurled into the water amidst a tangle of broken shrouds, stays, and splintering timbers.

There *Genii* remained upside down for a while, lodged in a rocky gully between the ledge and the cliff on the shore. Its black bottom pointed skywards, the scar made by the initial impact plainly visible as the waves continued to pound the hull mercilessly, the decks and superstructure grinding heavily against the unforgiving rocky shore, slowly wearing them away and scattering timbers and iron fastenings along the shore. Finally the wreck came to rest in a gulch in the cliff. All that remained was an empty upturned hull, its anchors and chains lying out to the southeast, as if still set. The shore in each direction was a mass of broken timbers. There were no survivors.

Altogether eleven men lost their lives aboard the *Genii* that night. The only man who did survive was the pilot, Captain James Clark, who had chosen not to return to the barque before the storm. While he sipped his rum in the safety of Mick Haggerty's place, the harbour's three-hundred-foot (ninety-metre) breakwater was entirely demolished, along with eight houses, a two-storey warehouse, a wharf, and a mill railway. But few people in New River were taking stock that night or in the days that followed of anything other than the terrible loss of life.

∼∼∼

In Saint John the wind had died at about noon on Monday, October 4, the skies cleared, and the sun shone brightly for a while. But the wind began to pick up again around 2 p.m. and continued to strengthen all afternoon. Wispy high-level cirrus clouds approached from the south and southwest, soon followed by denser lower level clouds, until by four the sun was completely obscured. About 5 p.m. the wind was blowing clouds of dust around in the streets, much to the annoyance of pedestrians. The barometer fell slowly and steadily all day, as it had been doing for the previous five days—there's nothing unusual in that—but as the storm approached the mercury began to fall more quickly. About 6 p.m. a torrential downpour began to soak the city streets, the mercury began to fall even more rapidly, and an hour later the hurricane hit.

One of the most unnerving features of any hurricane is the impossible noise of the wind. It doesn't howl. It doesn't even roar. It shrieks. It's a scary enough sound in broad daylight, but in pitch darkness it's enough to terrify even the bravest of souls.

And there was not even the slightest glimmer of light on the night of Monday October 4, 1869. No stars were visible, and the new moon was dark. The sky was heavy with scudding cloud, and sheets of rain were blowing sideways. In the city, people in their homes could at least take some comfort from the company of their families. But not James E. Earle.

Earle must have been the loneliest person in Saint John that night. He had only begun his job of keeper of the Beacon Light at Sand Point, at the mouth of Saint John harbour, three months earlier, after the previous keeper had quit his post. The lighthouse was almost brand new, built by the Dominion government at enormous cost to replace one that had been destroyed by fire in 1867, but was nevertheless, according to the *Saint John Morning Freeman,* "a miserable temporary affair." About 6 p.m. the weather was becoming extremely wild, the strong southwest winds driving large waves against the lighthouse, and Earle decided to head for shore. At low tide it was possible to walk to the light along the spit of sand that joined it to the mainland, but as the rising tide now covered the sandbar, he hopped into his skiff and started to row.

He did not get far, however, before he realized the sea was too rough for his little boat, and he reluctantly returned to the light tower. The wind and waves continued to increase all evening, and soon waves were breaking right over the tower, stripping off the shingles and boards from the lower storey, and washing away the guard rail, steps, boat, and ballast. The terrified Earle climbed to the top of the tower and closed the hatchway to keep himself and the lamp dry. But the winds had shattered the lamp's glass globe. The flames, fanned by the wind, set fire to the structure's woodwork, threatening to engulf the entire building. Earle managed to douse the flames and put a new globe over the lantern flame, then remained precariously at his post until about 4:30 in the morning, when he finally dragged himself ashore.

Beacon Light, Saint John, New Brunswick, at low and high tide.

While Earle was maintaining his lonely vigil, the storm was creating another peril that was even more destructive than the wind. When very strong winds blow for hours on end across a large area of sea they drive the surface waters ahead of them, piling them up into a storm surge. What's more, the low air pressure that accompanies a hurricane also pulls the surface of the ocean upwards, piling it even higher. While a hurricane remains over the ocean, deeper currents feed water away from the surface, but as it approaches shallower coastal waters, there no longer are any deep currents, and the water piles higher and higher.

All evening the tide rose steadily, but the surge was pushing it higher and faster than usual. Low water in Saint John had been in mid-afternoon, and high tide was predicted for 10:30 p.m. It was soon obvious that, with steadily strengthening winds blowing a storm surge directly into the mouth of the harbour, the wharves and slips were in danger of flooding. At 8 p.m. the barometer fell to its lowest level—29.3 inches (992 mb)—and the wind was at its wildest. By nine o'clock the water in the harbour had already reached the level of a typical spring high tide—and it still had an hour-and-a-half to rise before its predicted peak. Every few minutes immense waves rolled into the harbour, like nothing anyone had seen before.

The western side of the town, the suburb then known as Carleton (now West Saint John), was exposed to the full force of the wind and the storm surge that it was driving forward. Knowing how exposed the area was to the sea, a number of cautious families whose houses were nearest to the shore had moved to a safer place to wait out the storm. Others who had stayed put were now alarmed to find that not only were waves breaking right over their houses, but they were also hurling gravel and stones—even boulders—clear over their rooftops. For two hours straight the massive waves swept ashore. Houses were swept away or turned right around on their foundations—one even wound up in the middle of the street—and the contents of many were thrown out onto the pavement. Many were totally wrecked. As their occupants fled for their lives, barely escaping with the clothing on their backs, the air was filled with the cries and terrified screams of men, women, and children desperately trying to save whatever belongings they could. Fish houses were carried away, and not a stick was left standing where the wharves had been. Amazingly, amidst all this destruction, no lives were lost.

In the outer part of the harbour, everything was also utter devastation, as the sea rushed inward like a mill race, waves now following one another in rapid succession. By now all the wharves here and many of the surrounding warehouses were flooded. In the *Empress* warehouse, men desperately removed goods to safety as the water rose above the floor, but thought better of it when the building started to rock. They escaped only in the nick of time, because shortly after they left a large part of the building collapsed. There was also much concern in Saint John about the *Empress* itself. The vessel sailed a regular schedule between Saint John and Digby, Nova Scotia, and had been due to set sail from Digby that afternoon. Fears only grew worse when word spread that the *Empress* had left Digby as usual at 3:30 that afternoon. What no one on this side of the bay knew was that once out at sea, Captain Steen, a man known for his prudence, had second thoughts, and decided that the weather was becoming too nasty for the forty-nautical-mile (seventy-three-kilometre) passage. To avoid any risk to his vessel and passengers, he turned the *Empress* around and headed back to wait out the storm in the shelter of the port.

In the streets it was hardly any safer, as potentially lethal volleys of roof slates and shingles filled the air. In other warehouses men worked through the night to save their goods, and from time to time some of these buildings too began to shake on their foundations as they were moved by the surging waves. Some men gave up the effort, just like those in the *Empress* warehouse, preferring to take their chances at home rather than face the danger of being trapped inside a building that might collapse at any moment. Outside, waves rolling in from the bay became ever higher, crashing into the wharves all around the harbour with tremendous force. On the Market Slip, the scene was bedlam as heavy vessels tied alongside were tossed about like corks. They rolled from side to side so violently that it seemed to those aboard as though they would roll right over onto the wharves. The men's yells could be heard even above the roar of the wind as they did what they could to fend their vessels off. As the storm progressed, the conditions only got worse. At the South Market wharf, a schooner lying with

Vessels loading and unloading cargo at the Market Slip, Saint John, New Brunswick.

its bow to the waves was pitching so violently that it looked as though at any moment it would take a dive headfirst to the harbour floor.

At about 8:30 p.m. two vessels that were tied up alongside the Custom House Wharf, *Twilight* and *Ansell*, broke loose and were driven down upon another vessel, the *Armanella*, at Lawton's Wharf. The waves washed over the wharf and rolled up as far as Water Street. The *L. L. Wadsworth* and *Maude Potter* also broke loose and were sent flying around the harbour like toy boats.

At the height of the storm there was some slight relief from the terror of the pitch darkness as thoughtful residents in the vicinity of Lower Cove placed lighted candles in their windows. Others turned out with torches and did what they could to preserve life and property. By now rising water had heaved up many of the wharves. Others were torn up completely by the waves. The Anchor Line Wharf, Ballast Wharf, Lower Cove Wharves, Reed's Point Wharf, Blizard's Lumber Wharf, Maxwell's Lumber Wharf, Rankin's Wharf, Market Slip Wharf, Fairweather's Wharf, North Market Wharf, and South Market Wharf were all badly damaged or destroyed. The whole outer end of the *Empress* wharf was carried away. Only the most heavily built wharves, like Walker's Wharf, survived.

In Courtenay Bay, to the east of Saint John, James Van Horn and his family fled their home expecting at any moment that it would be swept away. John Wilson was not so lucky. His wharf, fish house, and its contents—his nets and gear, and his entire summer's catch of gaspereaux and shad—were destroyed when a wharf from a nearby shipyard was torn loose and smashed into them, leaving not so much as a stump to show where they had stood. The following morning, along with all the lumber floating in the bay and piled up on the shore, and amid the debris of wharves and shipyards destroyed by the storm, the tideline was strewn with a stinking mass of Wilson's rotting fish.

No one actually measured the wind strength on Monday night, but one report from Apohaqui, (near Sussex), 27 miles (45 km) northeast of Saint John, noted that an eight-inch board was torn from a shed and driven into the ground—to a depth of 2.5 feet (0.75 metres).

∼∼∼

On the Saint John River, the two steamers *Fawn* and *Olive* that plied their way between Saint John and Fredericton were forced to ride out the storm at anchor off Gagetown, but before the

Fishing shacks in Saint John Harbour.

Olive reached its anchorage, part of its superstructure was blown off. More than fifty houses and barns between Saint John and Fredericton lost their roofs or were completely destroyed. At Fredericton itself the damage was less severe, but serious for all that. Everywhere trees were uprooted or their trunks snapped, and their limbs were scattered all over the city. A reporter described the scene:

> *The Officer's Square is a mass of limbs and leaves, and every street has its record of the wreck. In Northumberland Street a new building in course of erection, was hurled bodily into the road and dashed to splinters; the roof of the car shed was lifted a portion of its length and blown away; barns and wood sheds were unroofed and demolished, and scarce a house in Fredericton but rocked to its very foundation. In the suburbs the trees were felled by the keen edge of the wind in hundreds...On the road*

to Hartt's Mills, or Oromocto, it is safe to say that every third building is a ruin, and in one place we saw where a house was unroofed, and the bed of the little children, from whence they were just snatched in time by their parents, occupied by the fallen bricks of the chimney. Every fence along the road is down or damaged and the hay is strewn about the fields as though it had never felt the scythe.

~~~

As the wind picked up during Monday, many vessels, large and small, were still at sea and began to run for shelter. But there was really nowhere to run. Many of the havens where they usually sought safety offered little protection in a storm as ferocious as this. One vessel that was known to be at sea during the afternoon, but which had not been sighted since, was the pilot boat *Lightning*, with two pilots and two or three boys aboard. Even though the pilots were known to be exceptionally skilled seamen, in Saint John their families and friends began to fear the worst. What exactly the *Lightning* had been doing is not clear, but as the gale picked up on Monday afternoon the pilots made a dash for Westport, the tiny harbour on Brier Island across the bay on the Nova Scotia shore. They were not alone. Altogether some fifteen vessels tried to ride out the storm at Westport.

By midnight the wind and rain were finally beginning to subside at the lower end of the Bay of Fundy, but the storm was by no means over yet. If anything, it was just about to unleash an even greater terror.

CHAPTER 7

Storm Surge

There is no end to the mystery of the tides.

—Hilaire Belloc

A chance encounter in the atmosphere off the coast of New England had already rekindled the fading storm's fire, and now another meeting—this time with the mighty tides of the Bay of Fundy—was about to add to its malevolence. As the people of Saint John had just discovered, when hurricane force winds get behind a rising tide, the results can be disastrous.

Stephen Saxby was mistaken in thinking the moon had a hand in creating this raging storm that had bullied its way north from the Bahamas and slammed ashore in the vicinity of Eastport and Saint John, but the moon did have a role to play now that the storm had entered the bay. For the moon exerts a powerful gravitational pull on the oceans—it is the main force that drives the tides.

Fundy is justly famous for its giant tides. At the head of the bay the tidal range—the difference in the water level between successive high and low tides—is greater than anywhere else in the world. The rise and fall are quite awe-inspiring, and the numbers are truly staggering. About a hundred billion tonnes of water move in and out of the bay twice every day. The sheer weight of all this water actually pushes down the earth's surface as it rushes into the bay—parts of Nova Scotia adjoining the bay are measurably lower at high tide than at low. At Cape Split, where a narrow channel constricts the flow of the tide into the Minas Basin, the amount of water that passes with each tide is more than the combined flow of all the water in all the rivers of the world. Here the tidal stream can reach a speed of 9 knots (more than 16 km/h). The water can rise as much as 3.3 feet (1 metre) in only twenty minutes. And at high tide the water level at the head of the bay can be as much as 25 feet (7.5 metres) higher than it is at the mouth—a ship steaming up the bay is actually "climbing uphill."

Ocean waters rise and fall in a rhythm that is tied to the motion of the moon—the sun has a similar effect, but it is much smaller. As the moon orbits the earth, it pulls the surface of the oceans nearest it upwards into a bulge, which in mid-Atlantic is less than 3.3 feet (one metre) high. This bulge would move freely around the world, following the moon's progress, if there were no continents to stand in its way. But wherever the bulge meets land or shallow water its motion is slowed and it is forced upward even more.

By the time the crest of an average Atlantic tide has made its way across the continental shelf and reached Halifax, Nova Scotia, it has increased in height only slightly, to about five feet (1.5 metres). But when the crest of that same wave enters the Bay of Fundy, the influence of the land becomes ever more important.

The bay is funnel shaped, broad at its mouth and narrowing toward its head near the Minas Basin, and this has given rise to a common misconception. It's sometimes claimed that the extremely high tides in the bay are partly due to the fact that as the water passes between the converging shorelines it has nowhere to go but upward. This is not so. The extreme range of the tides here is entirely to do with the bay's dimensions.

As it happens, these dimensions (or to be precise, the dimensions of the Bay of Fundy and the Gulf of Maine combined) give the bay a natural rhythm that is in tune with the rhythm of the tides—the bay and its tides are said to *resonate.* Perhaps the best way of explaining this is to imagine the water in a bathtub. If you start the water rocking by giving it a gentle push at one end of the tub, a wave will begin to pass to the other end. In a six-foot tub the wave will travel from end to end in about two seconds. This is known as the tub's *period of oscillation,* and it depends only on the length of the tub and its depth. If you now give the wave a gentle push each time it reaches one end of the tub or the other, it will soon increase in size, until eventually water starts sloshing over the rim and onto the bathroom floor. In the same way, a child's swing has its own period of oscillation, the exact time depending on the length of the ropes that support it. Once in motion, if the swing is given a gentle push when it reaches the end of its travel, it will ascend to ever-greater heights.

If there were no tides in the Bay of Fundy, the time it would take a wave to travel the length of the bay from its mouth to its head and down again—its period of oscillation—would be about thirteen hours. This is very close to the twelve hours and twenty-five minutes of the period between one high tide and the next. So, like the water in the bathtub, when the crest of the oceanic tide passes the mouth of the bay, it is given just the right push needed to greatly increase the slosh at the far end of the "bathtub," in the upper reaches of the bay.

By the time it has reached Yarmouth, the wave has dramatically increased in height to 12 feet (3.7 metres) and by the time it has reached Saint John and Digby it is 25 feet (7.6 metres) high. But it is still only getting warmed up, for by the time it reaches the head of the bay at Minas Basis (where the average rise and fall is 40 feet (12 metres), it sometimes exceeds 52.5 feet (16 metres).

As well as their twice daily tempo, the tides have other rhythms, too. At full and new moon the earth, moon, and sun are all more or less arranged in a straight line (sometimes with both sun and moon on the same side of the earth, and at others with them on opposite sides of the earth), and then they pull *together* to produce higher high tides and lower low tides than average. These are known as *spring tides* (the term has nothing to do with the season). During the other quarters of the moon, the first and third quarters, when the sun and moon are not in alignment, they are pulling *in different directions*, and the tides are smaller than average, with lower high tides and higher low tides. These are called *neap tides*. This cycle repeats about every two weeks.

So far so good, but the moon's orbit around the earth has yet another characteristic that throws another rhythm into the mix. The moon's distance from the earth is constantly changing because its orbit is not circular, but elliptical—at some times it is closer to the earth than at others. When the moon is closest to the earth (at *perigee*) its pull is stronger than when it is farthest away (at *apogee*), and so, at perigee, which occurs about every four weeks, the tides are pulled even higher.

Every now and then a spring high tide will combine with a perigean tide to produce an exceptionally high tide. Normally even the highest spring perigean tide will not cause any concern, as harbours, wharves, dykes, and other coastal structures are built with these tides in mind, and these events are predicted with great accuracy, years ahead, by tidal experts. But it is a different matter when, on very rare occasions, a powerful storm with winds from the southwest coincides with a perigean spring tide in the Bay of Fundy.

When it does so, the storm surge that develops spells serious trouble for settlements at the head of the bay. And that is precisely what happened during Saxby's storm.

For centuries the powerful Fundy tides have sculpted the coastline of the bay. They have also shaped the lives of the people
who live there.

There's a rather odd feature that surprises many people when
they first see the Bay of Fundy for themselves. The vast amount
of water that moves to and fro the length of the bay twice a day
has a powerful scouring effect on the red Triassic sandstone that
outcrops here, and it turns the water a rich red colour, not unlike
the colour of canned tomato soup. A portion of this fine red sedi-
ment—more than 105 million cubic feet (3 million cubic metres)
of the stuff erodes into the Minas Basin every year—settles out
with every flood tide in the sea meadows that made the place
attractive to the early French settlers, who began arriving in the
seventeenth century.

These settlers, the Acadians, discovered that, over thousands
of years, the ebb and flow of the tides had bestowed upon their
adopted land a deep, rich soil that was far superior to the skimpy
layer of dirt that covers most of the rocky interior of Nova Scotia.
It was ideal for farming. The land was level and the settlers did
not need to exert themselves in the backbreaking work of clearing rocks and trees to make it workable. Instead they set about the
backbreaking work of building earthworks to keep the tides out
of the sea meadows (salt marshes) and thus reclaimed this land
from the sea.

The Acadians were very good at it. Their ancestors had been
doing the same for centuries in Europe, so the first arrivals in
the 1630s brought with them techniques of draining coastal land
originally devised to create salt pans—salt being a pricey com-
modity at the time. In the New World, their first efforts were
made at Belle Isle marsh near the settlement of Port Royal, where
they constructed dyke walls to hold back the sea, fitted with

sluices known as *aboiteaux* to allow the reclaimed land to drain when the tides receded. The soil could then be ploughed, and when a season or two of rains had leached out the salt, it was ready for cultivation as pasture or arable land—probably the finest in the Maritimes.

The dykes were an amazing feat of engineering, and they withstood the worst that the sea could throw at them for more than two centuries. But they had one weakness. They were constructed to stand only about a foot proud of the highest perigean spring tide that the Acadians had ever experienced. Generations of Acadians had lived their entire lives without ever seeing a tide rise higher than the top of their dykes.

∽∽∽

The first realization that the dykes were in trouble on the night of Monday October 4, came in the rivers around Digby, Nova Scotia, where the steamer *Empress* had returned rather hastily on Monday afternoon to shelter from the storm, and had spent the night at anchor in the almost-landlocked Annapolis Basin. About the same time as water started to break over the wharves at Saint John, the storm surge was also rushing through the narrow Digby Gut, past the anchored steamer, and into the Annapolis River. By early evening the people of Annapolis Royal found themselves wading knee-deep on Lower St. George Street. Further upriver, at Bridgetown, water was soon flowing over the dykes and flooding the land behind. In the Cornwallis marshes hundreds of startled cattle, sheep, and other livestock that had been left out to pasture were overcome by the floodwaters and drowned.

It was the first taste of things to come further up the Bay of Fundy. Between Digby Gut and the mouth of the Petitcodiac River lie 85 nautical miles (160 km) of open water, but it takes only twenty minutes or so for a tide to cover that distance, and

Bay of Fundy dykelands in settled weather.

this tide was pursued all the way by the hurricane and its storm surge.

The storm was at its height at the head of the bay just as the tide was reaching its peak. The dykes were no match for the storm surge here either and it barged relentlessly into every river, creek, and inlet. Once again, as at Cornwallis, the surge poured over the dykes and into the low-lying lands beyond, taking with it vessels that had been lying in the rivers and carrying them miles inland.

Water in the Petitcodiac and Memramcook rivers rose more than six feet (two metres) higher than had ever been seen before. Almost every wharf on the rivers was carried away. Large sections of the approaches to the large bridge across the Petitcodiac were washed away, and were driven underneath the railway bridge at Jonathan's Creek, taking that out too. Every bridge between Moncton and Hopewell was gone, and almost every other bridge in the vicinity suffered some sort of damage.

When the dykes were breached on the eastern side of Shepody Bay, Dorchester Station, located in the middle of a marsh, was instantly swept by a wall of water some say was fifteen feet (4.5

metres) high. The station master, John Trites, rang the bell of a locomotive in a desperate appeal for help and his family were trapped in their house as water surrounded the station. As it entered the house and furniture began to float, the family dashed upstairs. Although neighbours heard the ringing bell, there was no way for them to reach the house without grave risk to their own lives, and the family spent the night in a terrified huddle on the second storey. More than six miles (ten kilometres) of track on the Eastern Extension Railway was washed out.

At Moncton the wharves were devastated, but fortunately the water did not reach the most heavily populated parts of town. A Mrs. Tidd, who lived at the foot of Church Street, was alone in her house, asleep in bed, when she awoke to find that floodwater was five feet above her floor, but she was soon rescued by her neighbours. In Hall's Creek marsh, the floodwaters lifted stacks of hay and deposited them all over the place, and carried the carcasses of drowned cattle upriver toward the town.

To the southwest of the town, Upper Coverdale was the scene of one of the most tragic events of the storm. Jacob O'Brien, his wife, and their four children, one of them still an infant, lived in a small house in the village. They had gone to bed on Monday night and tried to sleep as the storm raged outside. No doubt they feared the wind might take off their roof or tear shingles from their walls, but the thought of a flood had probably not crossed their minds. Then, like many other families that night, they awoke to find that not only was their ground floor quickly filling up with water, but their house was also completely surrounded, with all means of reaching dry land cut off. O'Brien quickly decided that the only hope of escape for his family was some sort of float, so he gathered several pieces of fence rail drifting past the house and lashed them together a makeshift raft. He lifted his wife and children aboard, then climbed on himself and pushed the raft away from the house, trusting that the wind would blow them to safety. It was not to be. The wind blew the raft across the river, but before it reached the other side it broke apart and the family were tossed into the water. Only O'Brien and his wife survived. Their four little boys all drowned. There is another bitter twist to this tale: when the parents of the dead boys returned to their home the next day, they found that it was still standing.

There was another tragedy near Hillsborough, where the tide rose with extraordinary speed—about eight feet (2.5 metres) within ten minutes. The wave broke over the dykes of the Petitcodiac River at about 9:45 p.m. A young woman about sixteen years of age called Huldah Bray was out riding with a young man, a mail carrier called C. E. Chesley, in his horse-drawn buggy. They were crossing the Lake Road just as the wave approached. For reasons known only to herself, the terrified woman leapt from the comparative safety of the buggy. Chesley jumped after her and was able to grab her just as the wave hit them both, but they became engulfed in a swirl of floating timber and were separated. Chesley somehow managed to cling to a piece of flotsam and scrambled to the safety of dry land, but he had seen Huldah Bray alive for the last time.

The floodwater at Hillsborough caused tremendous damage at the Albert Manufacturing Company, and rushed up the main road, sweeping away everything in its path—bridges, stores, barns, haystacks, even a coal wharf with four hundred tons of coal—and drowning three horses. A large boat was thrown over a dyke, landing in the middle of a freshly cut field. David Dryden's house was saved only when a large chimney collapsed upon it, and the sheer weight of bricks and mortar kept the house from floating off. He and his family climbed to the attic, now under a foot of water, where they spent the night. A large area of marshland in the village became a lake.

Another story has passed into legend, about how the storm saved a marriage. A young bride had asked her new husband to build their home with the kitchen window facing north, as she particularly liked the view in that direction. No, he argued, a north-facing window would make their house too cold. Instead he built the house with the kitchen window facing south. It did not remain so for long. On the night of the storm, floodwater lifted the entire house from its foundations and turned it right around. The kitchen window now faced north, and the young lady was happy again. A true story? Perhaps. It's certainly true that many houses were lifted off their foundations by the flood and ended up in new locations.

∽∾∽

It had been a bountiful summer for the hundreds of farmers who made their living around the head of the bay on the marshy land of Cumberland and Westmorland counties. The meadows, now cropped to a short stubble, were dotted with countless stacks of hay, fodder that was either left on the ground or piled in stacks to be hauled away later. Some of the more prosperous farmers had built barns on the marshes to store their hay, and these were now filled to bursting. Every farm next to a marsh depended upon it for the fodder that would see its livestock through another long Nova Scotian winter, now only weeks away.

As they retired with their families for the night on Monday October 4, they had every reason to be grateful to nature. Although the night was warm and muggy, open windows were soon fastened and shutters closed against the torrential rain and rising wind. As the wind continued to strengthen, no doubt many turned out to gather up loose items around their properties, and to check on their animals. No one had the slightest idea of what was about to engulf them and their homes. What was there to worry about? Never in living memory had water risen above the top of the dykes that protected their land from invasion by the sea.

It was about 10:30 p.m. when the surge started to sweep over the dykes, although no one could actually see it coming on this pitch-dark night. In no time at all, the marshes were covered to a depth of one or two feet. And here too, as the dykes started to give way, a wave swept with great speed across the marshes, picking up and carrying away everything that lay in its path—fences and haystacks, even barns. Two fishing schooners in the Cumberland Basin were lifted bodily over a dyke and deposited three miles (five kilometres) inland in the Tantramar Marsh.

Near Amherst, for several days before the storm, men had been loading the schooner *Active* with a cargo of cooking stoves, built at Robb's Foundry in the town. As the storm worsened, the owner, Mr. Robb, dispatched four men to help its captain secure the vessel in Gordon's Brook, a branch of the LaPlanche River. By now, however, the wind was so strong and the tide rising so rapidly it was not possible to reach the vessel, and the men took shelter in a nearby hay barn. They hadn't been there long before floodwater started to rise around them in the building, and they decided to make a run for it, following a fence leading from the barn to higher ground at Fort Lawrence Ridge. It was while the terrified men were struggling to find their way in the pitch darkness that the dykes burst behind them, and a wall of water rushed toward them. The wave overtook them, knocked them off their feet, swept the fence away, and took the men with it. Two managed to grab hold of some poles and saved themselves after drifting for two or three miles, where they finally reached the higher ground. Their two comrades, Henry Colbourne and Norman Siddell, were not so lucky—the last that was heard of them was their desperate cries for help. Both men drowned. The body of Siddell was never found.

Another fellow, an old man named Steward, who had been cutting grass on the marsh at Minudie Point, made a habit of sleeping in a barn there at night. The floodwaters picked up the entire barn, and as it floated away it started to break up. Fearing that he would be crushed by the falling building he swam to a passing haystack, and climbed aboard. Later—presumably when the tide turned—the haystack began to drift toward the open sea. He was saved only when the haystack reached the top of a dyke and became stuck fast. The following morning, after a miserable night spent clinging for dear life to his makeshift life raft, he was spotted and rescued by boat. On the same marsh nine horses and fifteen oxen, all drowned, were later found lying in a heap.

Nova Scotia was cut off from the rest of Canada. The roads were impassable and the telegraph lines were down, along with just about everything else, between Amherst and Sackville. A reporter from the *New Brunswick Recorder* described the damage at Sackville:

> *We had a terrific gale here last night and the most tremendous tide ever known! The marshes are completely covered with water as far as the eye can see. The loss of cattle is unknown. Thousands of tons of hay have been destroyed, whole barns and their contents have floated for miles in some cases. The railway is completely torn up. The roads are impassable between here and Amherst. Their condition is perfectly awful. Horses, oxen, sheep and pigs have been drowned in great numbers, and are lying amid rubbish of every conceivable kind. There is scarcely a fence to be seen; all are swept away. The loss is incalculable.*

∞∞∞

At about the same time as the dykes on the Tantramar marsh were breached, the storm surge was also cresting in the Minas Basin. It flowed over almost every dyke in its unstoppable rush forward, sweeping sheep and horses off their feet and drowning them in its murky water, some of their carcasses drifting out to sea. At Wolfville the tide rose higher than ever known before, and some 1,300 acres (520 hectares) of dykeland were flooded near Grand Pre and Horton. Rail lines were torn up along a 19-mile (32-kilometre) stretch of the newly completed Windsor and Annapolis Railway between Kentville and Horton, and two railway bridges between Wolfville and Port Williams were swept away. Seawater rolled inland as far as Sheffield Mills.

In the town of Windsor, dykes everywhere were breached. At 11 p.m. the residents of Poverty Point, near Smith Island, were woken when the dykes gave way, and within minutes the lowlands that lay behind them were flooded. Farmers released sheep and cattle from their pens to fend for themselves rather than risk their certain drowning. Some twenty-five families rushed to the safety of the upper storeys of their houses as water poured into the lower floors, flooding them to a depth of eight feet. In town, the sea approached to within a few yards of the old burying ground, Water Street resembled a canal, and water in the vestry of the Baptist Church reached a height of seven feet. Somehow the Wellington dyke survived, but dykes gave way at Falmouth and Newport. As the surge entered the Shubenacadie River it destroyed shipyards at Maitland and smashed wharves, carrying some of them half a mile inland. It crossed the lower streets in the village and flooded shops. Near the headwaters at Shubenacadie the flood destroyed a chapel. And when the dykes breached at Truro, travellers found many roads flooded and impassable, and the only way over the marshlands was by boat.

∾∾∾

On the other side of Nova Scotia, the Atlantic coast, it was a very different story. In late September and early October the citizens of Halifax had been basking in an unusual spell of warm fall weather, despite a prediction in *Belcher's Farmer's Almanack* that this would be a period of high winds and considerable rain. The city was also still basking in the warm afterglow of a visit by Queen Victoria's son, the young Prince Arthur, who had carried out all the normal duties required of a royal visitor. He reviewed the city's troops and the fleet, attended a ball at Province House, and a picnic at Princes Lodge before heading on to Quebec and Montreal to continue his tour. The most urgent item the *Morning Chronicle* could come up with for its "News and Gossip" column on Monday October 4, was that "London has a plague of Daddy Long Legs"—but it did also print a warning, recalling Frederick Allison's letter of October 1 in the *Evening Express*: "Our readers will do well to remember the great storm predicted to occur tomorrow…and, according to Mr. Allison, the warmer the preceding weather the heavier the storm." Never mind, said *Belcher*. That almanac's readers could expect fine and settled weather from the first to the sixth of the month. As events unfolded, Allison and Belcher were both proved wrong.

The afternoon of October 3, had been more like one in July, but about midnight there were several heavy rain showers. When Monday dawned it was still unseasonably warm, as it had been in Saint John. A strong breeze was blowing from the south, and there were occasional squalls. Frederick Allison checked his barometer at intervals and noticed that it was falling steadily—but not dramatically—throughout the day. At 7 a.m. it read 29.961 inches (1,015 mb), and at 6 p.m. it read 29.846 inches (1,010 mb), the kind of fall to expect if a stiff breeze is on the way, but nothing worse. At 2 p.m. the temperature had soared to 74°F (23°C). Anyone strolling along the waterfront could see that Haligonians had taken notice of Allison's warning, as the harbour and wharves bustled with activity while those whose livelihoods depended on the water prepared for the worst.

Most captains moved their vessels, large and small alike, away from the wharves where they had been lying and anchored them in the harbour. Some stripped them of spars and sails to reduce windage. Vessels that remained alongside were secured with so many lines that walking down the docks became a game of hopscotch. And steamship captains ordered their boilers stoked—they wanted to keep a good head of steam in case they needed to make a run for it. The steam tugs *Hoover* and *Lion* were made ready to go to the assistance of other vessels at a moment's notice.

Ashore, merchants cleared away everything they could easily remove from the wharves, and heavier goods such as lumber they tied down with chains. They removed goods from the lower storeys of their waterfront warehouses and hauled them to the upper levels.

About 3 p.m. it began to rain in buckets, and by evening the wind had strengthened to a moderate gale, but as the time for the evening high tide approached the water rose only to within two or three feet of the tops of the wharves. When Allison checked his barometer again at midnight, it had fallen only to 29.555 inches (a drop from about 1,014 mb to about one thousand mb, fourteen mb in seventeen hours)—nothing to indicate a serious blow was on the way. Of the predicted storm there was no sign: some strong winds, but nothing out of the ordinary for early October. A few old fences blew down in the north end of the city, a few shingles were torn from roofs on Water Street, and a wharf and bathing and boating houses belonging to the Beechwood estate on the shores of the Bedford Basin were carried away. Other than this, there were no reports of damage.

As darkness enveloped the city, the editor of the Halifax *Morning Chronicle*—clearly disappointed with the non-appearance of a storm, and without anything even as interesting as a plague of daddy long legs to comment upon—sat down to cobble together his editorial for October 5:

> *The storm of Monday night was not a success in the city—did not come up to the expectations of the public. The wind blew strongly and the tide was somewhat higher than usual but, as far as we have learned, no damage whatever was done in the city or to the shipping in the harbor.*

Frederick Allison may well have shared in his disappointment. Most likely he was feeling a little red in the face now that his warning had seemingly proved groundless. If so, his embarrassment was short-lived, as news from the Bay of Fundy began to trickle in to the offices of the *Morning Chronicle*. The trickle soon became a steady flow of reports from communities all around the bay. All said the same thing. A furious storm and storm surge had devastated the dykelands.

Cape Breton had felt a moderate gale, but no storm. At Antigonish, a fire burned twenty-one houses to the ground, its flames most likely fanned by the gale. And at Yarmouth there was a gale too. It blew hard enough to overturn a stagecoach, but it was certainly no hurricane.

Life in Halifax returned to normal with barely a hiccup.

CHAPTER 8
Aftermath

As predicted the storm came—not a storm but a tempest, such as this Province has not seen in the memory of living man.
— The *New Brunswick Reporter*, Fredericton,
Friday, October 8, 1869

For days the newspapers were filled with detailed lists of storm damage. In the best journalistic tradition following any disaster, reporters in affected communities were dispatched to seek out the eldest inhabitants, who dutifully declared they had never seen such a storm in all their lives. "It was the most fearful tornado ever witnessed here," added a reporter for the Eastport *Sentinel* the following day, his enthusiasm getting the better of his meteorology.

When lightkeeper James Earle finally made his way ashore in the early hours of Tuesday morning, battered and bruised but otherwise none the worse for wear, he promptly quit his job. He had had enough of lighthouses, and nothing on earth would persuade him to return to the scene of his previous night's terror. Thousands of sightseers flocked to the Saint John waterfront to view the damage. They could see right through the structure of the Beacon Light, where Earle had spent the most miserable night of his life. Man and lighthouse were both totally gutted.

Saint John merchants whose warehouses had been damaged remained cautious, lest another high tide follow, but they need not have worried. Instead, at 5 p.m. on Tuesday the low tide was exceptionally low—fully thirty-five feet (eleven metres) lower than it had been at high tide on Monday night—so low in fact that boys were able to walk on the mud outside one of the wharves. This was to be expected. After a *positive* storm surge— an exceptionally high tide—an exceptionally low one usually follows. A diligent reporter in Saint John was sent out once again to track down that city's most elderly person, who swore that he had never seen a tide retreat this far either. Meanwhile wharves that were not beyond repair were patched up and made service-able, for the time being.

Earle's bosses found a replacement, a Mr. Smith, to tend the Beacon Light on Tuesday night, but he had to replace the lamp with a new one and spent much of the night trimming it. The light was out when the *Empress* finally steamed into the harbour from Digby, more than a day late, and so the resourceful Smith indicated the lighthouse's position for its captain by ringing the bell.

As other vessels began to head back to their home ports, they sailed or steamed across a seascape littered with enormous rafts of logs, drifting lumber, sections of wharves, and vessels that had barely survived the storm. The pilot boat *Lightning* had been lucky. Its anchors had dragged, but it was one of five vessels at Westport that had survived the storm unscathed. Ten others were not so lucky—they were driven ashore—and one foundered at its anchors. On the passage back to Saint John from Brier Island, *Lightning* came across the dismasted brigantine *Annie*. When the winds had subsided, the brigantine's crew managed to rig a jury mast from bits and pieces of broken spars and had sailed as far as Point Lepreau, where the *Lightning* put a line aboard and towed the stricken vessel to Saint John. Everyone on board—men, women, and children—had survived the ordeal, but they had gone without water for two days and were in a terrible state. Another brigantine, the *Phoebe Ellen*, the one that had been run down by two other ships while at anchor in West Quoddy, was also towed in. Its captain reported that he counted thirteen vessels in the bay on Tuesday morning, most of them either dismasted or otherwise in a wrecked condition. The dismasted brigs *M. T. Ellsworth* and *J. Morton*, both out of Windsor, were towed into Saint John on Thursday by the steam tug *Perry*.

Barely one week after the *Genii* met its terrible fate, the hulk was sold at auction with somewhat unseemly haste, for $1,300; less than one-tenth of its insured value. And as always amid tragedy and disaster there was someone who sought to profit from it. A man from Deer Island who found the upturned hull of the schooner *Rio* and towed it in to Saint John sent its owner a bill—for the tidy sum of $450.

At Eastport the *New York*'s rudder was soon repaired and the vessel was put back into service. Passengers who had endured the horrific voyage to and from Eastport held a meeting and passed a number of resolutions, moved by John Hegan, and seconded by Captain F. W. Doane, and presented them to Captain Winchester:

> *That, Whereas, In the opinion of the passengers, it was, under Divine Providence, greatly owing to the coolness, sound judgment, and admirable management of the master, the pilots, the engineers, and the other officers of the steamer, and the prompt manner in which their orders were executed by their assistants and crew, that we owe the preservation of our lives from the very imminent danger with which we were threatened.*

Therefore Resolved, unanimously, That we hereby tender our most earnest, grateful and heartfelt thanks to Captain E. B. Winchester, the pilots, the engineers, and the other officers of the steamer, their assistants and crew, for the exercise, by them, of coolness and courage, which, under the blessings of God, has resulted in our safety and preservation from what seemed inevitable destruction...It was Further Resolved, unanimously, That, from the fearful severity of the gale causing the loss of both the steamer's anchors, her rudder, and threatening the destruction of the entire steamer, we have to express our confidence in the strength and splendid sea qualities of the steamer, and the power of her engine, which, directed and controlled by the wisdom that was manifested in her government and management, enabled us to reach a place of safety, and calls upon us to manifest, as we hereby do, the hearty satisfaction we feel in the qualities of the steamer, as well as the judicious management of her officers and crew.

A beaming Captain Winchester gratefully accepted the praise.

On Wednesday the International Steamship Company's steamer *New Brunswick*—companion vessel to the *New York*—arrived from Boston, having holed up at Gloucester, Cape Ann, during the storm. On the passage from Eastport, passengers counted a number of dismasted schooners and a steamer with its smokestack missing. The brig *L. Warren*, which had encountered the storm between Halifax and Cape Sable, with heavy winds almost on the nose, finally limped into New York with its foretopmast missing.

All around the bay, ships that had been left high and dry by the storm were hauled off on the next high tide. Many small vessels, mostly schooners, lay far inland. At Memramcook, a team of men had to dig a large ditch, or canal, across the marsh to the point where a large vessel had been stranded, so it could be floated off.

On Thursday the mutilated body of William Lake from Parrsboro, Nova Scotia, a deckhand on the schooner *Renown*, was found washed ashore at Musquash, where the vessel had wrecked. And on Sunday another body, this one a boy also believed to be one of *Renown*'s crew, was found in the same place.

∽∾∽

In the countryside of Charlotte County, around Fredericton, private roads were blocked by so many felled trees that the occupants of properties had to cut their way out, as if through a primeval forest. Likewise at St. George the air was filled with the sound of axes, as men laboured to clear fallen trees. Indeed, the trees in this whole area were ravaged by the storm. In some places not a tree was left standing over an area of 200 or 250 acres (8,100 hectares), in what became known as the "Saxby Blow-Downs." In other places every tree was blown down in narrow swaths a quarter of a mile wide. Men clearing up the blow-downs reported that they could walk for ten miles at a time on fallen trees, without ever once stepping on the ground. The downed trees remained a fire hazard for many years after the storm. But the storm was also an unexpected boon to many, providing much-needed work clearing up the mess and repairing the damage. "At this season, and in these dull times, when so many men, with families dependent on their exertions for sustenance, are willing to work, the late gale has proved to be rather a godsend, as its effects promise a large amount of unexpected labour," observed the *Saint John Daily Telegraph and Morning Journal*.

Everywhere telegraph lines were down, and it was three days

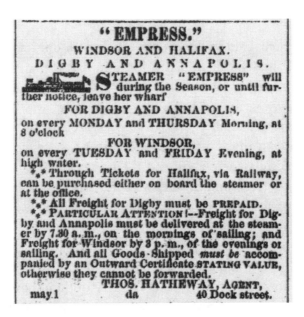

"EMPRESS."
WINDSOR AND HALIFAX.
DIGBY AND ANNAPOLIS.

STEAMER "EMPRESS" will during the Season, or until further notice, leave her wharf

FOR DIGBY AND ANNAPOLIS,
on every MONDAY and THURSDAY Morning, at 8 o'clock

FOR WINDSOR,
on every TUESDAY and FRIDAY Evening, at high water.

, Through Tickets for Halifax, via Railway, can be purchased either on board the steamer or at the office.

, All Freight for Digby must be PREPAID.

, PARTICULAR ATTENTION!—Freight for Digby and Annapolis must be delivered at the steamer by 7.30 a. m., on the mornings of sailing; and Freight for Windsor by 3 p. m., of the evenings of sailing. And all Goods Shipped *must be* accompanied by an Outward Certificate STATING VALUE, otherwise they cannot be forwarded.

THOS. HATHEWAY, AGENT,
may 1 da 40 Dock street.

Newspaper advert for the steamer Empress. *Its captain wisely returned to Digby to shelter from the storm.*

before communication was resumed between Saint John and the rest of the world. Between Saint John and New River alone, work crews removed forty-seven trees from the downed wires, and on the fifteen-mile (twenty-four-kilometre) stretch between New River and St. George, they removed a further 147. It was slow going, and crews could only clear a few miles of line in a day.

At Hillsborough, near Moncton, at the upper end of bay, the body of Huldah Bray was recovered, but her death cast a heavy pall over the village. The wretched bodies of the four O'Brien children were also found and laid to rest.

In the dykelands and marshes of the upper bay, thousands of tons of hay were scattered like chaff, and thousands of cattle, sheep, and horses were drowned. The marshes remained completely covered in floodwater for days, and it was a full week before all the dykelands had dried out, as each successive high tide was only a few inches lower than the one before, and the only means of going to and fro across the marshes was by rowboat. With the water finally gone, the once-green fields were now a muddy brown mush, often shrouded in a dismal fog and exuding a dreadful stench. Here and there the carcass of a dead cow or sheep lay bleaching in the sunshine. The shores also were littered with windrows of cattle and sheep. All the dykes between Minudie and Barronsfield were totally destroyed

Farmers at the head of the bay faced the daunting prospect of repairing or rebuilding mile upon mile of dykes. Dyke holders held meetings to decide what repairs to carry out and how to pay for them. At the first opportunity, some twenty-five to thirty men set to work repairing breaks in the dykes on the Tantramar Marshes, their work made somewhat less arduous by the spell of warm, dry weather that followed the storm, but, even so, they feared that they would be unable to finish repairs to some dykes before the next big tide. Some dykes, such as the town dyke at Avondale, were simply too costly to repair. And even when the dykes were rebuilt farmers were concerned that the land would take at least two years to recover. These concerns proved unfounded in some places as the land quickly returned to its fertile richness. In other places, it is said, it took three years. There were mountains of debris to remove; schooners and other vessels high and dry to be refloated; fences, buildings, and bridges to repair; hay sleds, wagons, and livestock to replace. In Cumberland County, fully one third of the hay crop was lost—but there was much that could be salvaged, too. In Amherst a meeting was held to decide how to distribute the salvaged hay, although there were still disputes over which hay belonged to whom. The storm "outweighs any calamity which has ever occurred," noted the *Amherst Daily News*.

Travelling was still either difficult or impossible in many

places, as roads remained impassable and railroad tracks awaited extensive repairs. The Windsor and Annapolis Railway, which had lost about eighteen miles (thirty-two km) of unconsolidated railbed, was eventually repaired, but the expense nearly brought the company to its knees.

In Maine the storm became known as the "Great Northeastern Rainstorm and Flood." Virginia, Maryland, Pennsylvania, New Jersey, New York, Connecticut, Massachusetts, Vermont, New Hampshire, and Maine all recorded rainfall of five inches or more (only Delaware and Rhode Island escaped the worst of the deluge), setting a record for rainfall in the mid-Atlantic and New England states.

No one knows for sure how many people lost their lives in the storm, and estimates vary. At least thirty-eight people were killed in the United States: fifteen in New York State, five in Pennsylvania, two in New Jersey, ten in Connecticut, and six in the other New England States. Most of the victims were drowned. Perhaps as many as seventy were lost in and around the Bay of Fundy. The loss of property has never been formally assessed but was certainly many millions of dollars.

The British public was angry. Saxby's predicted storm and tide had failed to appear—at least, in Britain—and people had missed out on their sport. They felt cheated. *Punch* magazine published a verse lampooning Saxby's prediction, comparing him with that most infamous astrologer and publisher of almanacs, Zadkiel:

THE TIDAL WAVE.
AIR—Something Handelian.
O SAXBY, thy prediction,
Concerning that High Tide,
To every man's conviction,
Last week was verified.
A Prophet thou so clever
Hast proved thyself to be,
Thy name shall live for ever
With ZADKIEL Tao-Sze.

Elsewhere in the same issue another paragraph appeared, *Punch's* formula for humour, as always, leaning on the groan-inducing pun:

The Seers at Sea.

ASK one of the weather (beaten) prophets what he thinks now of his prediction for the sixth of this month, and you will probably find him anxious to waive the question.

In the *Times,* weather correspondent R. H. Allnatt was prompted to write from Eastbourne, where the sea had been like a lake on the day of the expected storm, its surface barely stirred by a ripple, under a headline that was perhaps intended to parody and belittle Saxby's profession, "The Schoolmaster Abroad":

Of late years I have received charts and schemes, elaborate devices, deductions, and calculations from observers who believe in comets, the occult influence of planets, the power of the central fire, or volcanoes, or of the sun, or moon, or stars, and I have for

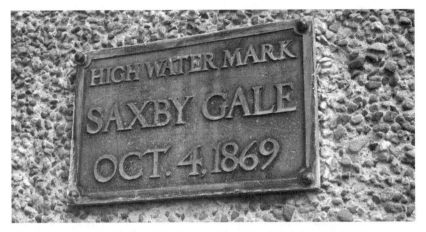

Plaque marking the height of the Saxby tide at the Tidal Bore Park, Moncton.

months had the curiosity to check off their diaries with the accomplished physical facts of nature, and I hereby declare that miserable prophets they all are. Like Murphy their great prototype...in nine cases out of ten they are all wide of the mark.

Punch was being excessively harsh on Saxby. To lump his name together with those of the astrologers Murphy and Zadkiel is entirely unfair. He was no astrologer. He was a man well-grounded in science who simply came to the wrong conclusion. Whereas astrology is based in nothing more than superstition, Saxby backed up his supposition that the moon influenced weather with a wealth of what he believed was supporting evidence—although it wasn't. The key word here is "evidence"—one that is not often mentioned in the same breath as astrology. Saxby imagined he was being scientific, but, like many who had gone before and many since, he made the mistake of inferring cause and effect where it did not exist, and without subjecting his "evidence" to stringent tests. The worst that can be said of him is that his science was wobbly.

As far away as Peru, Saxby's name was mud. His prediction

Grave marker of the four O'Brien children killed on the night of the storm.

had coincided with another, issued by a Professor Falb of Vienna, who also believed that the moon had a powerful influence over matters on earth. He had predicted an earthquake in South America for September 30 or October 1. In the coastal city of Callao, which lies only a metre or so above sea level, the warnings had caused its citizen to panic, and thousands fled to the countryside to take refuge. There were violent tremors and the sea did rise to unprecedented levels in northern Peru, but there was no catastrophe. On returning home, the citizens burned both Falb and Saxby in effigy.

ᴘᴘᴘ

In Canada, reaction to Saxby's prediction was instant and mixed. Just two days after the storm the *Saint John Daily Telegraph and Morning Journal* was lavishing praise on Saxby. "The storm of Monday night…was produced by such a conjunction of the Moon, Earth and the Sun as seem fitted to prove disturbing causes both as respects the tides and the wind…Even to persons who are not deeply versed in astronomy or meteorology it will appear that there are reasonable grounds for apprehended atmospheric disturbance mentioned by Mr. Saxby."

Nine days after the storm, however, Saxby's plausibility was being called into question by the rather more astute editor of the Saint John *Daily Morning News*, Edward Willis. After ascertaining that there had been no other major storms reported from anywhere else in the world on October 5, Willis concluded:

> *Upon the whole, although it is too soon to decide irrevocably upon Lieutenant [sic] Saxby's prophetic claims, the verdict upon the facts as they at present appear, must be, at last, not proven. We are not at all liable to be blown about by every wind of doctrine, scientific or sensational; but when sufficient evidence has been produced to warrant such a conclusion, we shall be glad to believe that Lieutenant Saxby's philosophy rests on a sound scientific basis and that his prophetic warnings were inspired by the most*

benevolent feelings, ands that the precautions taken
saved both life and property. At present, we can only
say that of his benevolent intentions there can be no
doubt, that the precautions taken at the season were
wise and profitable, but that the prophecy does not
seem to have been exactly or substantially fulfilled.

The *Saint John Morning Freeman* was equally dismissive: "…in regard to the storm itself," wrote the editor, "it may be remarked, that anyone, taking the whole world for a field of action, may with certainty, almost, predict high tides and heavy gales during the first week of October, for this period is peculiarly a time of storms somewhere or other…we think we are warranted in saying that the storm of the fourth was not occasioned by the cause as pointed out by Lt. Saxby."

More than forty years later, in 1911, the historian D. L. Hutchinson wrote: "The chances of a hit in Saxby's case were greatly in his favour. He had the whole world to range over, for he expressly stated the two hemispheres would be affected alike. If then in any locality on the surface of the earth a violent storm occurred on the day named, Saxby could claim a fulfillment of his prediction. In these latitudes during the month of October we are fairly certain to have gales, and that one of these should occur about the time indicated is nothing very remarkable."

What of Saxby's motives? It seems fair to say that they were honourable. He never profited from his predictions, and they brought him more scorn than plaudits. He clearly demonstrated a concern for the lives of seamen throughout his working life, but he came unstuck when his ambition to be taken seriously as a man of science outweighed his scientific acumen. And having struggled in vain to make a name for himself with his inventions, perhaps he saw the lunar theory as a way to finally gain the credit he felt he deserved.

Did his weather warnings save lives? Almost certainly. But there was a downside, too. When his storm warnings didn't materialize, they sometimes kept ships in port when it would have been perfectly safe for them to put to sea, hitting shipowners in the pocket unnecessarily. Weigh a modest financial loss against the possibility of lives being lost, and it's clear which way the balance should fall. But, as Robert FitzRoy had discovered, when forecast storms fail to appear, forecasters are perceived as crying wolf, and their subsequent forecasts carry less weight.

If Saxby's predictions did any serious harm it was to the efforts of FitzRoy and the other pioneers of forecasting, who were struggling to achieve their own credibility. Saxby was never taken seriously by the scientific establishment, but his predictions, and the astrological nonsense spouted by Zadkiel and others, were constant thorns in the side to the work of men like Fitzroy, because people always want to know the impossible: what the weather will be like—not just tomorrow or the day after, but also next month, next summer, or next winter.

∞∞∞

After his departure from Sheerness, Saxby initially found time on his hands, and used it to pen a new round of articles for the *Nautical Magazine*. No longer did they address the subject of weather. In a letter to the editor, Saxby advocated the use of India-rubber springs on ships' chain cables to prevent their breaking under shock loadings. He also took what might have been intended as a swipe at his previous employer, in a lengthy diatribe titled "Coals Used in Steamers," pointing out the enormous amount of coal that was being wasted in the Royal Navy's steamships.

By 1871 the Saxby family were living at Broxbourne in Hertfordshire, and Saxby had found another position that suited his talents, teaching navigation and seamanship to yacht owners

By 1881, when this cartoon appeared in Punch *magazine, weather forecasts had resumed in Britain, but they still provided rich fodder for satire.*

and merchant officers studying for their Board of Trade exams at the British and Foreign Sailors' Institute, on Mercers Street, Shadwell, London, presumably making a daily commute by steam train. In 1874 he published his last book, *The Yachtsman's Manual and Sea Officer's Guide*, in which he avoided the topic of weather altogether. But he had not abandoned his lunar theory. A number of newspapers continued to print his weather predictions, which also appeared annually in *Saxby's Weather Table and Almanac,* until his death. Stephen Martin Saxby spent his last years at Wormley, also in Hertfordshire, and passed away on March 11, 1883, aged seventy-eight. The *Times* did not print an obituary.

It was nine years after Saxby's enforced retirement before the Royal Navy tried again to improve the training of its engineer officers. In 1878 an engineering school was established aboard the screw-driven ship HMS *Marlborough,* followed by the shore-based Keyham College two years later. Executive officer training was improved by the opening of the Royal Naval College at Greenwich in 1873.

∞∞∞

The idea that the moon is somehow responsible for disturbances in the weather did not die with Stephen Saxby. Just two years after his death, in 1885, author Walter Lord Browne, like Saxby, claimed to be the first to propound the theory of lunar influence on weather disturbances. "Storms, tempests, cyclones," he wrote in *The Moon and the Weather: the Probability of Lunar Influence Reconsidered,* "are the results of the moon's action, and these are the most remarkable phenomena that first present themselves to us, and they are distinctly traceable to her influence, and all conditions of weather depend on them."

Even today a few ardent dissidents cling to the notion. Perhaps the best known is Ken Ring, a New Zealander who describes himself as a mathematician and long-range weather forecaster in his book, *Predicting the Weather by the Moon,* published in 2000. He puts out annual almanacs with weather predictions for New Zealand, Australia, and Ireland, and also maintains a website, "Predict Weather." As in the mid-1800s, no reputable scientist today would support his suggestion that the moon's motion upsets the balance of the atmosphere or causes weather disturbances. Ring claims he can predict not only weather but also earthquakes—an art that any seismologist will also tell you is quite impossible at the present time.

∾∾∾

Saxby's storm took place at a pivotal moment in the development of weather forecasting. In the years that followed, forecasting took a new lease on life. In 1870 the U.S. Congress passed a resolution that gave the U.S. Army's signals service the responsibility of providing "notice on the northern Lakes and on the seacoast by magnetic telegraph and marine signals of the approach and force of storms." In January of the following year, Cleveland Abbe, who had started issuing his weather "probabilities" from the Cincinnati Observatory just a month before Saxby's storm, was given the task of issuing forecasts—still called probabilities—three times a day. By 1872 his bureau's storm predictions covered the whole country. Within a few years "probabilities" became "indications," and in 1880 the term "forecasts" was finally adopted. In 1890 the work of the weather service was handed over to a new section of the Department of Agriculture, the U.S. Weather Bureau.

In England weather forecasts were not revived until April 1879, fourteen years after the tragic death of their creator, Robert FitzRoy.

Saxby's storm also led to renewed calls for storm warnings in Nova Scotia and New Brunswick. One of the most eloquent appeared in the *Saint John Daily Telegraph and Morning Journal* just one week after the storm:

> *Now is the time that the great necessity for storm telegrams under the management of the government is most urgently felt, when we hear daily reports of, and see with our own eyes, the unprecedented ravages*

of the storm. Had telegrams been promptly sent from different parts stating the hours of its arrival, the direction and force of the wind, appearance of the barometer, etc, pretty accurate calculations could have been formed of the progress and power of the storm, and the hour when it should reach any given locality. The telegram received by the captain of the steamer New York, from the agent in Boston, is an instance of the usefulness of such a system, and had this warning not been sent, and the vessel been overtaken before reaching a harbour of refuge, she would doubtless have fared much worse than on Monday night.

Frederick Allison and other amateur meteorologists had been arguing the need for storm warnings, of course, for years. When George Templeman Kingston was starting to establish his weather network in Canada in the early 1870s, Allison had been a natural choice to become his observer in Halifax. He was soon to become one of the network's key players in Atlantic Canada. Kingston granted Allison a unique status within his observing program, as Chief Meteorological Agent for Nova Scotia, and Allison recruited about twenty other observers throughout the province. Kingston appointed a man called Gilbert Murdoch as superintendent at Saint John.

Initially Kingston relied on Washington to supply storm warnings for Canadian ports, but another devastating hurricane in 1873 prompted the government to issue him with sufficient funds for a system modelled on the one FitzRoy had established in England, using drums and baskets (and lights at night) hoisted on masts at ports and harbours. In 1873, thirty-three of these "Drum Stations" appeared between the Great Lakes and the Atlantic coast. At first the warnings were often late and inaccurate, but they soon improved. Allison was never really happy that the warnings could be issued only from Kingston's central office—he was the man on the ground, after all—and the subsequent disagreements between Kingston and Allison can be added to the lengthy catalogue of disputes that mar the history of meteorology. Allison went to his grave in 1879, the disagreements with Kingston still unresolved.

Kingston's system, like that of Cleveland Abbe in the United States, was not initially able to provide warnings of Atlantic hurricanes. When Abbe issued his weather "probabilities" for October 3 and 4, 1869, he knew nothing about the hurricane that was drenching the eastern seaboard with a rainfall of biblical proportions. His telegraph network was designed to give warning of storms crossing the United States from west to east. There was no way he could have any forewarning of hurricanes approaching the continent from the West Indies.

But others were starting to take a keen interest in the possibility of forecasting such storms. The first to do so was not an American, but a Cuban, a remarkable priest called Father Benito Vines. Cubans long had a special interest in advance warning of these monsters—their island is frequently battered by Atlantic hurricanes—and when Vines became director of the Observatory at Belen College in Havana in 1870, he combed through records of past storms and began to make detailed observations of the daily weather. He quickly revealed an uncanny ability to predict hurricanes. Within five years he had established a network of observers on the island, his success soon becoming legendary, gaining him widespread recognition and admiration throughout the Caribbean region, and earning him the title "Father Hurricane."

By the 1890s telegraph cables had been laid between the U.S. and weather observatories in the Bahamas and Caribbean, and the U.S. Weather Bureau established a number of Caribbean weather observation stations in 1898, the same year President William McKinley proposed that the Weather Bureau should set up a hurricane warning network. Sadly, Father Vines's success, and that of the Cuban meteorologists who followed him, had aroused so much jealousy among his counterparts in the United States that eventually Willis Moore, who became head of the bureau in 1895, actually banned his staff from using Vines's forecasts. This petulant and high-handed action was partly to blame for the lack of a proper warning in 1900, when the deadliest hurricane ever to hit America almost completely destroyed the city of Galveston, killing at least six thousand people and perhaps as many as eight thousand.

In the early years of the twentieth century, hurricane warnings were issued from the Weather Bureau's Miami office. With the invention of radio telegraphy in 1902, ships at sea were eventually able to transmit reports of hurricanes. And as newer technologies evolved—instrument-laden weather balloons, radiosondes, weather radar, satellites, aircraft that could fly right into the eye of a hurricane, and powerful computers—meteorologists' understanding of these tremendous storms improved, as did their ability to forecast them. In 1967 the Weather Bureau's Miami office became the National Hurricane Center. In 1945 the U.S. military began identifying Atlantic hurricanes using women's names, and in 1950 the Weather Bureau began the practice that continues today of using both men's and women's names. The names of especially destructive or deadly storms—such as Andrew, Hugo, Katrina, and others—are dropped from lists, which otherwise repeat on a six-year cycle.

Until the mid 1980s, Canadian forecasters relied on the United States for warnings of tropical storms, but after Hurricane Gloria failed to deliver its predicted punch in the Maritimes in September 1985 there were calls for the creation of Canada's own hurricane warning system. In August 1987 the Canadian Hurricane Centre was established at Bedford, Nova Scotia, to track all tropical cyclones that pose a threat to Canada and to issue warnings of their approach. There was originally another centre on the west coast in Vancouver, but in 2000 the operation was centralized in Dartmouth, Nova Scotia.

The good news for the Maritimes is that if there is ever another Saxby the forecasters at the Canadian Hurricane Centre will certainly provide a timely warning of its approach. The bad news is that there almost certainly will be another Saxby.

CHAPTER 9

Why Saxby Still Matters

Unfortunately, as with earthquakes, there is a marked diminishment in perception of threat with time.

—George Parkes, 1997

In the first week of October 2002, the telephones at the Meteorological Service of Canada in Dartmouth, Nova Scotia, began ringing off the hook, as anxious residents of the communities around the Bay of Fundy desperately tried to contact a weather expert. The reason for their alarm was an article in the *Canadian Farmers' Almanac*, pointing out that on Sunday, October 6, the same astronomical and tidal conditions that prevailed during the Saxby Gale—a new moon, accompanied by a perigean spring tide—would line up again. The almanac was also predicting heavy rains and gale-force winds for the period. Was the combination of high tide and storm that made Saxby so dangerous and destructive about to be repeated?

If the scare proved one thing, it was that the long-term weather predictions of today's almanacs are about as reliable as they were in 1869. As things turned out, it was a false alarm. There was no catastrophic storm surge. Yet anxiety still haunts those whose homes and livelihoods—even their lives—remain vulnerable to the very real threat should a high perigean spring tide ever again coincide with a powerful storm in the Bay of Fundy. Could a storm like Saxby's happen again? What can be done to prevent it? And how prepared are the emergency services to deal with the situation?

∽∽∽

Saxby's storm was not the first hurricane to push an extreme tide over the top of the dykes in the Bay of Fundy—and the consensus among scientists is that it will not be the last. It is not a question of if there will be another Saxby, they say, but when.

On the night of November 3, 1759, one of the earliest hurricanes on record struck the Canadian Maritimes. It was probably at least as frenzied as the Saxby storm, and it delivered a similar storm surge. At Saint John the water rose six feet (1.8 metres) above the highest of normal high tides, and the terraces of Fort Frederick, tucked well inside the harbour, were inundated. And at Fort Cumberland (Fort Beausejour) the tide washed away seven hundred cords of firewood that had been stacked at least ten feet (three metres) above the height of the dyke that protected it. Perhaps it's idle to speculate on whether the Acadians ascribed the storm surge of 1759 to God's wrath for the appropriation of their dyked lands after their expulsion four years earlier. It is tempting, nevertheless!

Atlantic Canada is lucky that, by the time they reach our offshore waters, most hurricanes are either weakening or disintegrating altogether. Atlantic hurricanes are classified on what is known as the Saffir-Simpson scale, which grades them into one of five categories, the least powerful being Category 1 and the most powerful, Category 5 (see Appendix 1). No hurricanes of

Category 3, 4, or 5 have ever travelled this far north. Two powerful storms originally believed to be Category 3 reached the Canadian coast in the last 120 years, but they have since been downgraded: one ripped through St. Margarets Bay in Nova Scotia on August 22, 1893, and Hurricane Luis made landfall in Newfoundland on September 11, 1995.

There have been other hurricanes, like Saxby, that gain in strength, rather than decay, as they approach Canadian waters. At least two of them have much in common with the Saxby gale in this respect. The first was the "Great New England Hurricane" of 1938. In fact it was the first serious hurricane to come ashore in New England since Saxby. This Category 3 storm was moving so fast—its forward speed was clocked at sixty knots (110 km/h)—that it was able to keep its momentum when it passed over colder waters. In the United States, flash floods killed more than six hundred people, but Canada was spared on this occasion.

Hurricane Hazel, on October 15, 1954, was another. It was also a late-season hurricane that merged with another mid-Atlantic weather system as it travelled north. Hazel eventually journeyed inland without losing much of its power, dumping torrential rainfall—more than eight inches (two hundred mm) in twenty-four hours—in southern Ontario, causing massive flooding. Eighty-one people lost their lives, making it the most serious hurricane ever to hit central Canada. A similar kind of strengthening may also have occurred in September 1775, when a tropical cyclone blasted Newfoundland and the islands of St. Pierre and Miquelon, and again in August 1873 when a hurricane drove through the Gulf of St. Lawrence, taking some six hundred lives. A recent example is Hurricane Igor, in 2010, which joined forces with another low-pressure system and reintensified as it approached Newfoundland. Its impact was devastating.

It's not only hurricanes that blast the Bay of Fundy with winds vicious enough to create the same kind of storm surge as Saxby.

Satellite image of Hurricane Bob in August 1991. Saxby's storm would have looked very similar.

There are non-tropical winter storms that can easily match hurricanes in their sheer violence, and in the storm surges they generate.

The Groundhog Day Storm of February 2, 1976, for example, caused a storm surge of four feet (1.2 metres) at Saint John, and five feet (1.5 metres) at Yarmouth. It was a particularly nasty storm that moved very quickly and with little advance warning. On February 1 a gale warning had been issued, and it was upgraded to a severe storm warning overnight. There was a large storm surge in Maine—at Bangor it was 10.5 feet (3.2 metres) above the predicted tide. It arrived at the mouth of the Bay of Fundy during a period of spring tides, but fortunately they were only moderate springs, not perigean. Had the moon been at perigee the results would have been catastrophic in Saint John. Even so, its effects were bad enough. Ships broke loose from their moorings, the storm flooded large areas near the mouth of the bay, and there was severe coastal erosion. As gale-force winds slammed into the area they tore down power lines and severed communication links, and the damage to boats, wharves, and buildings was estimated in the tens of millions of dollars. Southwest Nova Scotia caught the worst of it, especially the coast between Barrington Passage and Digby. In Yarmouth residents witnessed the highest tide in thirty years.

Another winter storm, on January 10, 1997, a deep, northward-moving depression, passed just to the west of Saint John on the track that spells danger of a storm surge in the bay, and sure enough it produced a surge of two feet (0.67 metres) at Saint John. This storm did coincide with a perigean spring tide, and gave the highest tide seen in Saint John for fifty years. And when the so-called Storm of the Century on March 14, 1993 passed west of the Maritimes, it caused storm surges at Yarmouth and Saint John of just over three feet (one metre), but luckily the tides were neither high nor in the spring part of their cycle.

More recently there have been a number of near misses—hurricanes that looked as though they would take the same track as Saxby but altered course at the last minute. Between 1885 and 1996 nine hurricanes passed just to the west of the Bay of Fundy, or directly through it, and there were four near misses. If any one of these had coincided with a run of perigean spring tides it could have caused flooding on the scale of Saxby's storm.

Hurricane Ginny, on October 29, 1963, was one of these near misses. Another late-season storm, it took a rather curious track, making a loop-the-loop off the coast of the Carolinas before heading northeast and hitting southwest Nova Scotia, where its path followed more or less the south shore of the Bay of Fundy. As a result, New Brunswick caught the worst of the storm's rainfall, and Halifax caught the worst of the winds, but the bay was spared from a storm surge. More recent hurricanes that looked like they would follow a similar track to Saxby include Earl on September 4, 2010, which forecasters early on thought would track up the Fundy, but instead turned east and rolled into Nova Scotia's St. Margarets Bay. And in late August 2011, forecasters originally thought Hurricane Irene was headed the same way, before it eventually went further west, making a final landfall near Brooklyn, New York.

In the years following the Saxby Gale and its attendant storm surge, the event became embedded in the collective memory of the inhabitants of the Bay of Fundy. Stories were passed on, mostly by word of mouth, from generation to generation, but hardly anyone in the world of science took particular notice of this storm, despite —or perhaps because of—the sheer rarity of such events. But by the turn of the twentieth century, both historians and scientists were starting to take an interest.

The first was New Brunswick historian and naturalist William Ganong. In 1911 Ganong wrote to Cleveland Abbe, who by now had been chief of the U.S. Weather Bureau for forty years, asking for information about the storm. Abbe was quite unaware of this particular storm—which seems strange considering the flood damage it caused in the United States—but he wrote to the London Meteorological Office, asking for information. The London office sent him some news items gleaned from the *Halifax Chronicle* and the *Halifax Daily Express*, and a few sketchy biographical details of Stephen Saxby. A contemporary of Ganong's, New Brunswick historian D. L. Hutchinson, wrote the first "scientific" account of the storm, also in 1911, in which he traced the storm's path fairly accurately, and rather perceptively observed that it was probably of tropical or sub-tropical origin.

The first attempt to establish just how high the storm surge had risen on the night of Saxby was made the year following the storm by engineers who were carrying out a survey at the head of the bay, but their method inadvertently created a myth that persists to this day about the range of the bay's tides. It was all due to their choice of a reference point from which to measure the rise and fall.

There had been a proposal to build a canal—the Baie Verte Canal—across the narrow, marshy, fifteen-mile (twenty-five-kilometre)-wide Chignecto isthmus, the neck of land that is Nova Scotia's umbilical cord to the rest of Canada, to make a shortcut for shipping from the Bay of Fundy to the Northumberland Strait and the Gulf of St. Lawrence beyond. In the same way that land maps need a datum from which to measure the elevation of hills and valleys, so do maps and charts need a datum from which to measure the depth of water and the heights of tides. Datums are always arbitrary, and the surveyors working on this project chose a datum one hundred feet below the level of the top of the dykes, to allow for the huge difference in ranges on opposite sides of the isthmus, and to keep their calculations neat and tidy, without the complication of negative values. The canal was never built, but since that survey several people mistakenly based the height of the Saxby tide on this datum. The Saxby tide of course rose higher than the top of the dykes, which led to greatly exaggerated readings of a tidal range of a hundred feet (30.48 metres) or more. In fact, the extreme range is little more than half this.

The British Admiralty began recording the rise and fall of the tides on a tide gauge in Halifax Harbour as early as 1851, but the navy kept records only for a few years, and predictions made from them proved wildly inaccurate. It was not until 1890 that the Canadian government established the Tidal and Current Survey and appointed a talented engineer, Dr. William Bell Dawson, as superintendent. Bell Dawson became the first person to make a proper study of the Bay of Fundy tides, and he made a very thorough job of it. One of his first tasks was to set proper bench-marks and datums and, as the nineteenth century gave way to the twentieth, his pioneering work began to reveal some of the secrets of the bay's tidal patterns. One of the difficulties he encountered was that in many parts of the bay it is not possible to install tide gauges that can measure the full tidal range. The range of the tides at the head of the bay is so extreme that many sandbars and inlets dry out completely at low water, making it impossible to measure water levels throughout the whole rise and fall. This makes it difficult to interpret some of the tides' intricacies.

Bell Dawson discovered one especially curious facet of the Fundy tides. In the bay, the range of the tides is affected more by the constantly changing *distance* between the moon and the earth (remember, the moon's orbit is not circular but elliptical), than the moon's *position* in relation to the earth (which is visible as its phase—full moon, new moon, and so on). The high tides increase in height significantly when the moon's orbit brings it closest to the earth (at *perigee*) and decrease when it is farthest away from the earth (at *apogee*). This effect is so strong that it can override the normally pronounced variation between springs and neap tides (the spring/neap cycle). So a perigean neap tide in the bay of Fundy can be higher than an apogean spring tide. This is quite unlike tidal rhythms found elsewhere, and is another aspect of the Fundy tides—beyond their remarkable range and volume—that makes them unique.

In the summer of 1916, Bell Dawson visited Burntcoat Head, in Nova Scotia's Minas Basin, to measure the range of the tides. Burntcoat Head is a rather unprepossessing place. From the land-ward side it doesn't even look like a headland, and if you blink as you drive past, you'll miss it altogether. But it has long held a significance for sailors. A lighthouse was built here in 1858 on a piece of land that eventually became separated from the main-land by the powerful scouring action of the Fundy waters. Early in the twentieth century it was dismantled and rebuilt on the mainland, where its beacon guided mariners for almost another seventy years. Today the area is occupied by a small park, and a replica lighthouse stands on the site. But Bell Dawson's visit gave the place a very special claim to fame. The average tidal range here (the difference in height from one high water to the next low water) is 47.5 feet (14.6 metres), and its highest range is 55.75 feet (17 metres). There is a greater tidal range here than anywhere else in the world.

Bell Dawson also talked to the inhabitants of Noel Bay, a few miles east of Burntcoat Head, who were able to show him exactly how high the Saxby tide had risen on the ruins of an old house on the banks of the Noel River, which still bore the tideline. From this he was able to make an estimate of just how high the tide had risen at 1 a.m. on October 5, 1869, the night of Saxby's storm.

∝∝∝

In the 1960s, almost a century after Saxby's storm hit the Maritimes, a team of scientists working in support of the Apollo space program found themselves in an unlikely setting—deep within dusty archives, poring over bound volumes and micro-film of old newspapers, personal diaries, town records, and ships' logbooks. What on earth could this work have to do with a space program?

The scientists were weather experts who had been given the task of providing accurate forecasts of Atlantic hurricane tracks to the space program, which, in view of the location of NASA's space center—Cape Canaveral, Florida—is not really surprising. NASA needed this information when planning and executing its launching and landing schedules. They realized there was a great deal to be learned about modern hurricanes by trawling through records of historical storms. For decades, weather observers, amateur and professional, ashore and afloat, had been keeping such records during most of these hurricanes: wind speed and direction, wind shifts, peaks and lulls, cloud cover, and rainfall— as far back as Saxby's storm and beyond—and that was the data the scientists were after.

Like William Redfield, Matthew Maury, and Benito Vines, who had undertaken similar ventures in the nineteenth century, they found the painstaking work time-consuming and tedious— but it paid dividends. Although much of the information they retrieved was at best sketchy, the scientists were able to determine the intensity and what they call the "best track"—some say it is really a "best guess"—of several hundred historic Atlantic hurricanes. It's a fairly straightforward matter to arrive at a hurricane's approximate track from studying successive accounts of its progress, but to arrive at estimates of wind strength, in the absence of detailed records, is a little more difficult.

If scientists can't get at the data that they are looking for directly, they will usually find a more devious way. If the barometric pressure at the centre of a storm, its geographical location, and its approximate extent or radius are all known, there is a rule of thumb for estimating its wind speed. For example, in the region of the North American coast between latitudes thirty-five and forty-five degrees North, a storm with a central pressure of 28.9 inches (980 mb) will have maximum sustained surface winds of 73 knots (135 km/h), and if the pressure falls to 28.4 inches (960 mb) the winds will increase to 90 knots (167 km/h).

The scientists collected a vast amount of data, which they assembled into a database called HURDAT, simply an acronym for "hurricane data." It showed the position and intensity of every known cyclone from back as far as 1886, at six-hour intervals throughout its life.

The database was not without its faults, however, and in the mid-1990s a pair of Cuban scientists, José Fernández-Partagás and Henry F. Díaz, used additional sources to update and extend the information. They looked at a whole range of further records, from newspaper shipping reports to government documents, and were able to add information about the tracks of hurricanes as far back as 1851.

The database was again revised in the early years of this millennium, and an astonishing amount of information about past hurricanes is now available to anyone with access to the Internet, at the click of a mouse—complete with coloured maps showing the tracks and intensities of every known tropical cyclone (including hurricanes and tropical storms).

As for Saxby, the scientists are able to say with a fair degree of confidence that on the evening of October 4, 1869, although the hurricane had yet to make its landfall, when its effects were being felt in Massachusetts, the storm had a central pressure of 28.5 inches (965 mb), a radius of maximum winds of 30 nautical miles (55 km), and was packing wind speeds of 80 knots (150 km/h), making it a Category 1 hurricane on the Saffir-Simpson scale. By the time it made landfall just a few hours later it had intensified into a Category 2 storm with sustained winds of up to 90 knots (167 km/h). If anything, they say, these figures are conservative, so the actual wind speeds were most likely even higher.

It was not until the mid- to late-1990s that another group of scientists, this time in Halifax, Nova Scotia, got together and began to take a close look at the Saxby storm. They realized this was a kind of "benchmark" storm, one that set a standard against which other storms in the region could be evaluated. By studying historical hurricanes like this one, as well as other storms like the Groundhog Day storm of 1976, they hoped to discover what triggered such events, enabling then to predict similar events in future, to predict what kind of storms would cause what kind of surge, and to predict the extent of flooding to expect from them. The *ad hoc* group included meteorologists Jim Abraham and George Parkes, tidal expert Charlie O'Reilly, and geologists John Shaw and Alan Ruffman, and their conclusions were a stark reminder to the inhabitants of the Bay of Fundy about the risks of coastal living.

Building on William Bell Dawson's work, the scientists were able to calculate that the surge in Saxby's storm caused water to rise at least 5.6 feet (1.7 metres) higher than the highest astronomical tide—that is, the highest tide that could have been predicted from just the positions of the sun, moon, and earth in relation to each other—and possibly as high as 6.9 feet (2.1 metres).

They also concluded that what made Saxby's storm so devastating in the dykelands in the upper part of the Bay of Fundy was an extremely rare set of circumstances. A hurricane—it could have been any other kind of intense storm—travelling rapidly toward the north, and arriving at the mouth of the bay at the exact same time as one of the highest possible high tides. The storm then tracked inland, west of Saint John, so that its strongest winds drove a storm surge the full length of the bay and piled it up at the head.

Hurricanes are always accompanied by a storm surge, which is frequently more destructive than the wind. Storm surge, as Saxby demonstrated, is powerful stuff. And Saxby was *almost* a worst-case scenario.

But bad as it was, Saxby's storm could have been worse. The high spring perigean tide of October 5, 1869, had not quite reached the peak of its cycle. The highest spring tides always occur about two days after a full or new moon, and the moon had been new on October 5. If the hurricane had arrived two or three days later, it would have chased an even higher perigean tide into the Bay of Fundy, and the storm surge would have been an

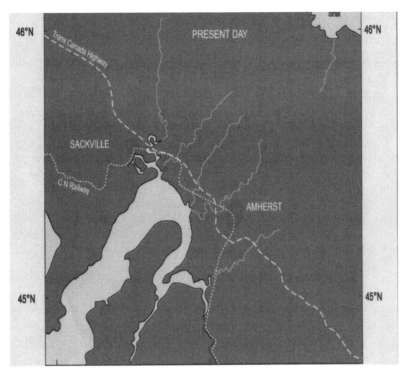

The level of a normal high tide in the Upper Bay of Fundy.

additional two feet (0.6 metres) higher.

Terribly destructive events like the Saxby storm are so rare in the Bay of Fundy—and this may seem counter-intuitive— because the enormous rise and fall of the tides in the bay actually *reduces* the risk of storm surge damage. The explanation is really quite simple. In places where the range of the tides is relatively small—as it is along the coast of Florida, and at Halifax on the Atlantic coast of Nova Scotia, where the maximum range is about six feet (two metres)—a surge of six feet (two metres) will cause water levels to rise above the normal level of the highest tides at *any stage* of the daily tidal cycle. In other words, in these places, such a storm surge *always* spells trouble. But at the head of the Bay of Fundy, where the tidal range can regularly reach an enormous fifty-two feet (sixteen metres), it is only at—or very close

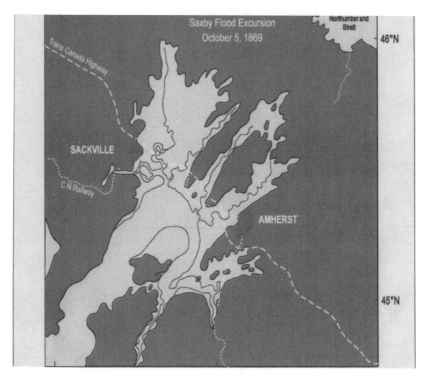

The area that was likely flooded during Saxby's storm.

to—high tides that a six-foot (two-metre) surge will cause flooding, and this can only happen during a *very limited period* in any day. At other times, such a surge on top of a tide of, say, forty feet (twelve metres) will reach a height of only forty-six feet (fourteen metres), still well below the level that would overtop the dykes. What's more, we know when these tides will occur, as they are predicted with great accuracy years in advance.

In other words, in the Bay of Fundy there is only a very narrow window of opportunity for a storm surge to become a serious flood threat, and by keeping an eye on weather forecasts and tide tables, more or less anyone can see when the right set of conditions exist. But this should not lull anyone into a false sense of security. One of the conclusions that the scientists came to was that there is every chance that a similar set of circumstances will arise again. And that the probability is increasing with the passage of time.

One of the most startling results of this project is a map, compiled by marine geologist John Shaw, in which he reconstructed the extent of flooding that probably occurred during the Saxby tide, based on the estimated storm surge—in effect, the areas that would once again be flooded if such a storm were to reoccur.

He was using a technology known as LIDAR, based on very sensitive laser-based instruments, which when mounted in an aircraft can be used to survey floodplains and measure elevations with great accuracy. With computer technology it is then possible to use the survey data to produce maps that show the area that will be flooded by a storm surge of, say, three feet (one metre) or six feet (two metres). It's a tool that is used extensively by governments in flood-prone regions. Simulations for Nova Scotia show that one of the main streets in Truro, for instance, could be flooded to a depth that would submerge parked cars and flood houses to the top of their first-floor windows. On the outskirts of the town, near Highway 102, there is an area that locals call "fast-food alley," where a number of eateries and a Canadian Tire store are built on reclaimed land: it also would be inundated.

Dykelands also support the only road and rail links between Nova Scotia and the rest of Canada, and these would again be severed—the CN railway line is built on top of a dyke, with the TransCanada Highway just beyond. The Radio Canada installation on the Tantramar Marsh is also vulnerable. The towns of Amherst and Sackville, both bordering the Tantramar, can expect serious flooding, as can Moncton and thousands of acres of surrounding farmland.

~~~

There is one final aspect of the Bay of Fundy tides that should give

The Tantramar Marsh at high tide. A storm surge like the one during Saxby's storm would rise some 6 feet (2 metres) higher.

cause for concern. In addition to the various tidal rhythms mentioned earlier in this book, there is another that makes tides higher at some times than at others. It is a much longer cycle, lasting 18.6 years, and has to do with another complicated aspect of the moon's orbit, known rather dauntingly in the language of astronomy as "precession of the lunar elliptic plane." What happens is this: The plane of the moon's orbit around the earth crosses the plane of the earth's orbit around the sun twice a month, and due to the physical phenomenon known as precession, each time the two planes intersect at a point (or "node"), that point has moved a little further westward from the time before. It actually takes 27.2122 days to complete a full circle. Every 18.6 years, or after about 250 cycles, the point of intersection returns to precisely the same place. When it does so, there is a particularly high peak in the tides. What this means for the Bay of Fundy is that every 18.6

years the highest tides reach another crest, this one superimposed on the spring and perigean rhythms. And 9.3 years after that, the cycle is at a point where the highest tides are significantly lower. The difference between the highest and lowest tides produced by this particular aspect of the moon's influence amounts to about two feet (0.6 metres). The next peak of this cycle will occur in September 2014.

Of all the extraordinary changes in the weather in recent years perhaps the most alarming for anyone living in the firing line of Atlantic hurricanes is a perception that these storms are becoming more frequent. This seems to be real enough, but what is not so clear is whether it is part of a long-term trend.

Some of the reasons we are experiencing more hurricanes now than a few years ago are well known. Scientists over the last half century or more have come to realize that the oceans and the atmosphere are so closely knit that they are often now regarded, not as separate entities, but as one. It also has become clear that weather patterns are intimately linked to patterns found in the oceans, and there are some links of which scientists are certain. One of these is a cycle of alternating warm and cool temperatures in the surface waters of the Atlantic Ocean, caused by the ocean's circulation. It goes by the name of "the Atlantic Multi-decadal Oscillation," or AMO. In the Atlantic, warm surface water from the equatorial region migrates north, and when it reaches Arctic waters it cools, sinks, and flows southward again at great depth. The AMO is part of a bigger, worldwide circulation known as the Ocean Conveyor Belt, which goes in fits and starts, and in turn causes the temperature of the ocean surface to fluctuate. The fluctuation between highs and lows is only about 0.4°C—but, remember, sea surface temperature is critical to a developing hurricane.

The AMO was discovered only in 1994, but temperature records going back some 130 years or so show that warm periods match very closely with periods when the greatest number of Atlantic hurricanes have occurred.

From the 1940s to the 1960s, when the ocean was in a warm phase, there was a lot of hurricane activity; and from the 1970s to the early 1990s, when the ocean was in a cool phase, there was less, especially in eastern Canada. Atlantic hurricane activity peaks about every twenty years, and since the mid-nineties we have been in the warm part of the cycle. This is expected to last until at least 2015, and maybe longer. Since 1995, only two or three hurricane seasons have been quieter than normal. The year 2009 was particularly quiet, for instance, but the explanation for this anomaly possibly involves yet another interplay between ocean and atmosphere.

It is another cycle of oceanic warming and cooling that scientists are quite certain has a powerful influence on hurricane activity. Called El Niño, it is a warming of the ocean that occurs about every five years near the equator in the Pacific Ocean, and is usually followed by a cooling period known as La Niña. The whole cycle causes changes in the atmosphere known as the El Niño/La Niña Southern Oscillation, or ENSO, and it has many worldwide impacts on the weather, most of them serious, including floods and droughts.

As for hurricanes, in El Niño years the winds in the Atlantic and Caribbean have more shear—that is, they blow more unevenly at different altitudes, and this hinders hurricane formation. It has been well established that in El Niño years these atmospheric conditions will prevent several hurricanes from forming.

It's often claimed that climate change and global warming are playing a role in increasing the frequency of hurricanes, but this remains uncertain. There is some evidence to suggest it is at play, but its effect is masked by the stronger influence of the AMO and is very difficult to detect. If it is having an effect, says Chris Fogarty, a forecaster with the Canadian Hurricane Centre, that effect is minimal. It might seem only natural that oceans warmed by man-made changes in the atmosphere would provide more fuel for hurricanes in the form of warm, moist air, but it's not as simple as that. Hurricane formation also depends on a *difference* in temperature between the upper and lower atmosphere, so it is important to know what effect global warming is having on this temperature gradient. If the upper atmosphere is warming faster than the lower atmosphere, thereby reducing the temperature difference, there would in fact be fewer and weaker hurricanes in the future. Some computer models show this to be the case, others don't, so there is no simple link, and scientists are still trying to figure it out.

In other words, climate change is not the smoking gun it is often made out to be, at least as far as hurricanes go. But as for a storm such as Saxby, Chris Fogarty agrees with other scientists who have studied the storm. He is certain one like it will happen again.

✧✧✧

The risk of damage from another Saxby has also increased since 1869, due to another change in the oceans. Sea level has risen in Atlantic Canada and continues to rise—slowly, but surely. There are three reasons for this.

First, several thousand years ago continental Canada was covered by an enormous sheet of ice, forming a hump one to 1.8 miles (2–3 km) thick over central parts of the country. The weight of all this ice pushed the earth's crust downward in these regions, and, at the same time, on the edge of the ice sheet in Atlantic Canada, where the ice was relatively thin, the crust was forced upwards. But then, about ten thousand years ago, as the ice cap melted and the weight was taken off, the crust in the middle of the continent began to rebound, and the crust in Atlantic

Canada began to subside. It has continued to do so, and recent measurements in Halifax Harbour, at the Bedford Institute of Oceanography, show that the crust is subsiding at a rate of 6.6 inches (17 centimetres) every century. And as a result, of course, sea level is rising by the same amount.

Second, sea level is also rising due to the water that is added to the oceans by melting glaciers in the Arctic (not by melting sea ice—that is already floating in the water). And, third, global warming—however much climate-change skeptics try to deny it—not only increases the amount of meltwater entering the oceans, but as water becomes warmer it expands and this, perhaps surprisingly, causes another significant rise in sea level—one equal to that added by the meltwaters.

The end result is that a tide gauge at Halifax shows sea level in Nova Scotia is rising at a rate of 1.1 feet (35 centimetres) per century. And as sea level rises it also causes an increase in the range of the tides. This means higher high tides as well. There are, of course, more people living near the coast of the Bay of Fundy than there were in 1869, and, as they are beginning to realize, rising sea level is a fact of life, a hazard that must be faced.

Those who shoulder responsibility for keeping the dykes in good shape, like Gerald Post, with Nova Scotia Agriculture, are well aware of both the challenges presented by that sea-level rise and that the future may well bring more frequent and more powerful storms. Even now, high winds at high tide will often cause minor flooding somewhere along the dykes. Exactly where depends on the wind direction. "A southwest wind," he says, "is not an issue at Grand Pré or Wolfville, for instance, but it is at Masstown, near Truro, or at Advocate Harbour. And there are parts of the bay that are affected by *any* strong wind."

Until the early 1930s, the dykes were managed and maintained by the farmers who owned the marshlands that lay within them, at their own expense; but in the Great Depression the market for hay collapsed, there was no longer enough money to maintain the dykes, and many fell into disrepair. This was addressed in the 1950s and 1960s by the federal government's Maritime Marshlands Rehabilitation Agency, which rebuilt the old dykes and built new ones, and then sold them—all 144 miles (240 km) of dykes and 261 aboiteaux (sluice gates that allow the marsh to drain itself of fresh water but prevent sea water from entering) to the Province of Nova Scotia for one dollar. Unfortunately, no one had been thinking about sea-level rise at this time, and the dykes had simply been rebuilt to the same height as before.

The structure of many dykes also remains as when they were built—first by the Acadians and then by the "Planters" in the seventeenth and eighteenth centuries. They were built almost entirely from sods cut from the marshes, sometimes laid over a foundation or cribwork of logs. They gain their strength from a covering of vegetation—million upon million of grass roots that bind the soil together—and this has to be carefully tended. It is mowed once or twice a year, the hay removed, and weeds controlled by Nova Scotia Agriculture, which employs a full-time staff of eight to maintain the dykes. Crews work year round, inspecting the dykes to ensure they remain firm against constant erosion from the sea. During the reconstruction of the 1960s rock was used, at least on the seaward face of the dykes, and major repairs are made with rock these days. Dykes also sometimes sag, and need to be topped up in height. The maintenance crews also keep a constant watch on the foreshore, the land outside the dykes that is covered and uncovered twice a day by the tides and protects them from wave action. Much of this land is constantly shifting, and sometimes is chewed off by the tidal currents and waves. If it erodes close to a dyke it has to be repaired, again with rock.

Despite these efforts, the dykes as they exist today would stand no chance against another Saxby. The Acadians, remember, built their dykes only a tad higher than the highest tides they had experienced. Today, the dykes are maintained at one foot (0.3 metres) above the highest predicted astronomical tide. Another Saxby could overtop them by 5.8 feet (1.8 metres) or more.

∼∼∼

What are people living in vulnerable areas to do? They can take measures to prevent or minimize flooding, or accept that it is inevitable and be prepared to deal with it when it happens. They can protect their communities by raising the dykes, raising buildings to better withstand flooding, and by planning future development to avoid further construction in high-risk areas. And they can prepare for the worst by planning a suitable emergency response to flooding.

The dykes around the Bay of Fundy have barely kept up with sea-level rise, and the only way to be sure of preventing future encroachment by rising sea level or storm surge is to build them much, much higher. This is not simply a matter of adding material on top of existing dykes; they must also be made wider if they are to retain their strength. No one has yet put a dollar figure on the cost of raising all of Nova Scotia's existing dykes, but it would certainly be prohibitively expensive for the Province without outside funding. And it would be only a temporary measure at best: within decades the dykes would need to be raised again. It will inevitably become a question of priorities.

Today Nova Scotia's dykes protect some 44,000 acres (17,800 hectares) of dykelands—that is, lands that would be covered by a high tide if the dykes did not exist. By and large, farmland can recover from flooding events even as large as Saxby, but flooding is a far more serious matter where people live. And the big difference between now and 150 years ago is that far more people live on these lands. In some places, significant parts of towns are now built on what was once agricultural marsh—houses, hospitals, schools, sewage plants, and so on. And in towns like Windsor,

A hand-built dyke. The top is only about 1 foot (30 cm) above the level of a typical spring high tide.

Truro, and Amherst, much of this development has occurred directly behind the dykes. A study carried out at Annapolis Royal suggests that it would be less expensive to build a levee to protect the town than to pay for the flood damage that would result if one is not built. Other municipalities, such as Kentville, are already coming to grips with the real risk of flooding and are taking practical measures, including an additional dyke and pumping stations, and planning new buildings at elevations well above possible flood levels.

Despite these efforts, the dykes offer no defence against a storm surge on the scale of Saxby, a fact that's well known to staff at Nova Scotia's Emergency Measures Office (EMO), the provincial government's organization charged with coordinating a rapid and effective response to protect people and property during such an emergency. It is their responsibility to ensure that individual municipalities have adequate plans in place to deal with such a situation.

The response begins as soon as forecasters with the Canadian

Hurricane Centre spot a hurricane that appears to be tracking toward the Maritimes—any hurricane, not just one that might arrive along with an exceptionally high tide. They immediately alert EMO at its Joint Emergency Operations Centre in Dartmouth. "The good thing about tropical storms like Saxby is that we know quite a few days ahead that they're coming," says Bob Robichaud, a warning preparedness meteorologist with Environment Canada. His role is to provide as much warning as possible of an approaching hurricane and to constantly update information throughout the event. These days, forecasters use a number of computer models that give a sense of the storm's track and intensity and the way it is behaving, and they usually know five days or more ahead if a storm is likely to hit the Maritimes. The forecasters are in touch with EMO on a daily basis, but as the storm approaches communications with EMO become more frequent and intense.

"My role," says Dominic Fewer, Emergency Management Planning Officer for Nova Scotia's Central Zone, an area that includes flood-prone coastline of the Bay of Fundy, "is to put that weather information into a risk-analysis model, establish the potential worst-case scenario, and make sure that information is passed on to the municipalities, each of which has its own Emergency Management Coordinator, who is responsible for ensuring effective plans are in place, so they can make good decisions at ground level." The municipalities have hazard-specific plans in place to deal with incidents such as fires, floods, and other perils. A number of smaller municipalities have pooled resources with their neighbours to form more efficient and more effective regional emergency organizations. The Truro centre, for instance covers Stewiacke and Colchester County.

When the storm is still two days out, the meteorologists and EMO staff get together with the municipal coordinators to establish what measures are already in place and what kind of additional assistance they may need from the Province to help manage the situation. The top priority has to be evacuation of people from high-risk locations, but there are many other considerations, particularly maintaining essential services such as water, power, medical care and supplies, sewage treatment, food transportation routes, policing, and even banking—or quickly restoring them after an interruption. "People are a lot less resilient today than they were in 1869," notes Dominic Fewer, wryly.

The closest the Maritimes have come to a Saxby-like situation in recent years occurred on April 1, 2003, when extensive flooding across the southern Maritimes put EMO's plans to the test. In this case, the floods were not caused by a storm surge but by very heavy rainfall from an intense low pressure system, adding to rivers already swollen by the spring snow melt. As much as 4.3 inches (109 mm) of rain was recorded in Amherst, Nova Scotia, most of it during a twelve-hour period. The town of Oxford and the counties of Cumberland and Colchester all declared states of emergency. In Truro, which experiences flooding quite frequently, this was the worst event in thirty years. About 125 people were evacuated from their homes when the Salmon River burst its banks, and firefighters were kept busy rescuing more than thirty people who were trapped in their homes by the fast-rising waters. A further forty were evacuated in Kentville when the Cornwallis River flooded.

It was, perhaps, a taste of things to come.

Acknowledgements

In England I would like to thank Ben and Lila Fenton of Kew for their hospitality; William Spencer, military records specialist at the National Archives in Kew, who guided me through the tortuous channels of British Admiralty records; Garry Spence at the British Library, St. Pancras, for accessing a slew of back issues of the *Nautical Magazine*; staff at the Caird Library, National Maritime Museum, Greenwich, for help accessing documents; and Rebecca Watts, archive assistant at Gonville and Caius College Cambridge, for information.

In Canada I received help accessing documents, photos, and other materials from the staff at Dalhousie University's Killam Library; from the staff at Nova Scotia Archives and Record Management; from Lynda Silver, Librarian at the Maritime Museum of the Atlantic; and Christine Little of the Archives and Research Library, New Brunswick Museum. Don Alward from the Albert County Museum, New Brunswick, kindly provided the photo of Stephen Saxby. Nancy Saxby provided details of the Saxby family history. I thank them all.

Thanks also to Alan Ruffman for discussion and to retired tidal expert Charlie O'Reilly for patiently explaining the complexities of the Fundy tides. I am especially indebted to John Shaw of the Geological Survey of Canada for explaining many aspects of storm surge, and to Chris Fogarty of the Canadian Hurricane Centre for sharing his vast knowledge of hurricanes and for preparing the weather and hurricane track maps; both John and Chris kindly reviewed parts of the manuscript—any mistakes that remain are of course mine, not theirs. I thank Gerald Post of Nova Scotia Agriculture for information on the Fundy dykes, and Dominic Fewer of Nova Scotia's Emergency Management Office and Bob Robichaud of Environment Canada for explaining emergency preparedness procedures.

At Formac, thanks to publisher Jim Lorimer and production editor Martha Tuff for help in shaping the book, to Laurie Miller and Amanda Lucier for many useful suggestions and improvements to the text, and Meredith Bangay for design.

As always, I thank my wife Sheila Nunn for her constant support and encouragement.

Appendix 1

The Saffir-Simpson Scale of Hurricane Intensity

This is an abbreviated version of the scale. Wind speeds refer to speed averaged over a one-minute period in the United States. Elsewhere they refer to speed averaged over a ten-minute period.

Category 1: Minimal Damage
Wind speed 64–82 knots (118–153 km/h). Storm surge 4–5 feet (1.2–1.5 metres). Little damage to well-constructed buildings. Some coastal roads flooded. Minor damage to piers.

Category 2: Moderate Damage
Wind speed 83–95 knots (154–177 km/h). Storm surge 6–8 feet (1.8–2.5 metres). Damage to roofs, doors, and windows. Some trees blown down. Considerable damage to piers.

Category 3: Extensive Damage
Wind speed 96–113 knots (178–210 km/h). Storm surge 9–12 feet (2.7–3.7 metres). Structural damage to some residences. Foliage stripped from trees, large trees blown down. Serious coastal flooding. Smaller structures near the coast destroyed.

Category 4: Extreme Damage
Wind speed 114–135 knots (211–249 km/h). Storm surge 13–18 feet (4–5.5 metres). Extensive damage to buildings. Complete

roof failure on small residences. Trees blown down. Major damage to lower storeys of buildings near the shore. Some coastal buildings washed away.

Category 5: Catastrophic Damage

Wind speed greater than 135 knots (249 km/h). Storm surge greater than 18 feet (5.5 metres). Complete failure of roofs on many residences and industrial buildings. Extensive shattering of window glass. Some complete building failures. All trees blown down. Total destruction of structures near the shoreline.

Appendix 2

The Weather during Saxby's Storm

Weather Reports from the United States

October 4, 1869. Nantucket Island. At 3 p.m. the barometer fell to 28.70 inches (972 mb) when the wind veered from SE to SW (*Nantucket Enquirer*, as quoted by Ludlum, 1963). This was the lowest pressure reported during the storm, but was not at its centre, which was perhaps as low as 960 mb (HURDAT).

October 4, 1869. Gardiner (about 60 miles NE of Portland, Maine). Around 7 p.m. the barometer fell to 28.99 inches (982 mb) with wind veering from SE to SW (Ludlum, 1963).

The general direction of movement of Saxby was NNE.

Weather Records at Saint John

October 3	Barometer (in/Hg)	Temperature ($^\circ$F)
8 a.m.	30.120 (1020 mb)	57
2 p.m.	30.010 (1016 mb)	61
10 p.m.	30.005 (1016 mb)	59
October 4		
8 a.m.	29.923 (1013 mb)	63
10 a.m.	———	70
2 p.m.	29.780 (1008 mb)	70
6 p.m.	9.527 (1000 mb)	—
10 p.m.	29.332 (993 mb)	62

October 5
8 a.m.	29.450 (997 mb)	55
2 p.m.	29.456 (997 mb)	55
10 p.m.	29.665 (1004 mb)	46

5 p.m.	Wind increased to gale
6 p.m.	Rain began to fall
8:30 p.m.	Blowing a hurricane from S by E
9 p.m.	Wind reached maximum force, rain almost ceased
10 p.m.	Wind began to subside and shifted to SW

The barometer reading at 10 p.m. on October 4th was the lowest during the storm. The total rainfall on the 4th was 0.530 inches. (Source: Hutchinson, 1911, from the records kept by Gilbert Murdoch.)

The average air pressure at sea level is 29.92 inches of mercury, or 1013 mb (Sheets, 2001, p. 34). The lowest barometer reading ever recorded was 26.22 inches (888 mb) in 1988, during Hurricane Gilbert (Larson, 2000, p. 121).

There was a *New Moon* in Halifax and Fredericton at approx. 10 a.m. on October 5, 1869. The moon was also at *perigee* at 3 a.m. on October 5, 1869.

Notes

Introduction

page 15. *how vicious a storm*…With thirty-seven recorded deaths in the United States, Saxby ranks Number 220 in the National Hurricane Center's list of "Deadliest Atlantic Tropical Cyclones." That list does not take into account another seventy or so fatalities in Canada—apparently deaths outside the States do not warrant inclusion!

page 16. *more than two centuries*…It was unusual in 1869—but not unique—for a hurricane to be named after a person. There were rare precedents, such as Solano's Storm of 1780, named for a Spanish admiral who was forced to abort an attack on the British at Pensacola when a storm scattered his ships (Larson, 2000, p. 51). The system of naming hurricanes that is so familiar today was not introduced until 1950.

page 17. *a day or two ahead*…Sheets and Williams, 2001, p. 9.

Chapter 1: The Calm before the Storm

page 24. *peculiar-looking vessel*…The voyage of the *Thames* is described in detail in Kennedy, 1903.

page 25. *At Plymouth*…where it arrived on June 6, according to the *Times*.

page 25. *a ten-year-old boy*…"I remember having, upward of fifty years ago, seen the first steamer pass through the Downs," wrote Saxby in the *Nautical Magazine* (vol. 38, 1869, p. 402).

page 29. *estimated at nearly three times that*…Anderson, 2005, p. 57.

page 30. *on a third thunder*… Gomme, George Laurence and A.C. Bickley (eds.) *The Gentleman's Magazine Library: Being a Classified Collection of the Chief Contents of The* Gentleman's Magazine *From 1731 To 1868. Vol. 9, Biographical Notes.* London: Elliot Stock, 1889.

page 30. *Mars and Saturn*…Ibid.

page 31. *James Glaisher*…Anderson, 2005, p. 62.

page 32. *circulation of between twenty-two thousand and seventy thousand*…Farnell, Kim. "Everyone talks about the weather, but nobody does anything about it…" Online: www.skyscript.co.uk/meteorology.html (accessed December 7, 2011).

page 32. *for almost seventy years*…Morrison adopted the pen name Zadkiel after his hero, the seventeenth-century astrologer William Lilly, who used the same pseudonym.

page 32. *in poor health*…O'Byrne, William R. *A Naval Biographical Dictionary: Comprising the Life and Services of Every Living Officer in Her Majesty's Navy, from the Rank of Admiral of the Fleet to that of Lieutenant, Inclusive.* London:

John Murray, 1849. Online: www.archive.org/details/cu31924027921372 (accessed December 8. 2011). Also see the *Oxford Dictionary of National Biography* (ODNB).

page 34. *Royal Meteorological Society*...Walker, J. M., "The Meteorological Societies of London." *Weather*, November 1993. Vol. 48, no. 11, pp. 364–72. Online: http://www.rmets.org/pdf/metsoclondon.pdf (accessed December 8, 2011). Morrison also came to public attention in June 1863 when he sued Rear-Admiral Sir Edward Belcher for libel. Belcher had accused him of spreading superstition, charlatanism, and profiting from crystal gazing. Morrison won his case, but was awarded only nominal damages and no costs. (Anderson, 2005, p. 77). The motive that drove men like Morrison to publish their almanacs was certainly money, but was that all? It is hard to square Morrison's earlier selfless lifesaving effort with the work of a charlatan who placed self-interest above concern for his fellow human beings.

page 34. *to take them seriously*...Cox, 2002, p. 7.

page 34. *tongue in cheek*...Kittredge, George Lyman. *The Old Farmer and His Almanack: Being Some Observations on Life and Manners in New England a Hundred Years Ago, Suggested by Reading the Earlier Numbers of Mr. Robert B. Thomas's Farmer's Almanack*. Cambridge: Harvard University Press, 1920, p. 306.

page 35. *under that title in 1832*...*Belcher's Farmer's Almanac* had made its debut in 1824 as *The Farmer's Almanac*.

page 37. *the only failure*...*Cunnabell's Nova Scotia Almanac and Farmer's Manual* was first published in 1834 by Jacob Cunnabell, the brother of William Cunnabell, as the *Nova-Scotia Almanack* (which has weather predictions on its monthly pages), but was taken over by William in 1837. It was not quite as successful as *Belcher's* and was shorter-lived; it folded in 1861 after a fire at the printing press.

page 38. *threatens storms and showers*...For more see Pearce, 1864, chapter on the moon.

page 39. *barometric pressure*...Sir John Herschel, "The Weather and Weather Prophets," *Good Words*, Vol. 5, 1864, pp. 57–64.

page 42. *the* Clermont...Olmsted (1857), however, says Redfield became interested in 1822 in a steam-propelled boat that had been invented by a fellow townsman.

page 46. *enormous distances*...This was in 1853. Redfield briefly turned his attention to the study of tornadoes and found they have much in common with hurricanes, but soon turned back to his main interest, hurricanes.

page 47. *the standard work for mariners*...ODNB.

page 48. *quadrupled*...The force of the wind is proportional to the square of its velocity.

page 48. *forms a whirlpool*...Fleming, 1990.

Chapter 2: A Storm of Ideas

page 49. *village of Bonchurch*...Saxby probably moved here sometime around 1840.

page 50. *in the English Channel*...S. M. Saxby, letter to the London *Standard*, September 16, 1869.

page 50. *East India Company*...In his younger days Saxby may have been teaching, but he once mentions that he was "late of the East India Company." He also mentions being at sea off Madagascar, which seems to fit.

page 50. *a career in teaching*...In his curriculum vitae for the navy in 1858, Saxby notes that he has been filling the role of headmaster for thirty years, which suggests he had been teaching for ten to twelve years before he arrived at Bonchurch. His first four children were all born in Middlesex and Kent between 1831 and 1838.

page 51. *rich in fossils*...William Redfield, who shared Saxby's fondness for collecting fossils, would also have enjoyed the Isle of Wight. Both men had species named after them.

page 51. *for several miles*...Landslips continued occasionally until the early nineteenth century; one was recorded in 1799 and another in 1810.

page 52. *Hoploparia saxbyi*...*Annals and Magazine of Natural History*, 2nd series, Vol. 14, 1854 pp. 116–18 and Plate 4, Fig 1. A lot of Saxby's fossils ended up in the Woodwardian Collection in Cambridge's Sedgwick Museum. According to the Cambridge University Commission's *Report on the State, Discipline, Studies, and Revenues of the University and Colleges of Cambridge*, 1852, p. 118, S. M. Saxby Esq., of the Isle of Wight, was one of the benefactors of that great paleontological series.

page 52. *whitewash for centuries*...W. H. Davenport Adams. *The Garden Isle: The History, Topography, and Antiquities of the Isle of Wight*. London: Smith Elder & Co., 1856, p. 196 (available on Google Books).

page 52. *archery contest*...*Nautical Magazine*, Vol. 31, 1862, p. 364.

page 55. *fit his theory*...Cox, 2002, p. 37.

page 56. *evidence of the facts*...quoted in Ludlum, 1969.

page 56. *phenomena of this storm*...Mooney, 2007, p. 21.

page 59. *Coriolis effect*...Physicists still argue about whether it should be called the Coriolis force or the Coriolis effect. I have gone with "effect," because even if you believe it really is a force, it is indisputably an effect.

page 60. *and slightly inwards*...The popular myth that water draining from a sink rotates counterclockwise in the northern hemisphere and clockwise in the southern hemisphere is completely false—the Coriolis effect is way too slight to govern such a small-scale event.

page 60. *sun, moon, and stars*...Graney, Christopher M. "Coriolis Effect, Two Centuries Before Coriolis." *Physics Today* Vol. 64, Issue 8. August 2011, p. 8 (Online: http://dx.doi.org/10.1063/PT.3.1195).

page 62. *at sea of any kind*...Maury, who was passionate about the navy, was also highly critical of its management and organization. Writing under the pseudonym Harry Bluff, he penned a number of articles that came to the attention of the navy's top brass. Although annoyed by these diatribes, they were forced by public pressure to act on many of them. One of Maury's criticisms was the lack of a proper education establishment for the navy's midshipmen, a point that was in part responsible for the navy's decision to establish the U.S. Naval Academy at Annapolis, Maryland.

page 63. *tedium of the task*...Schlee, Susan. *A History of Oceanography: the Edge of an Unfamiliar World*. London: Robert Hale & Co., 1973.

page 64. *a "ten-year man"*...*The Biographical History of Gonville and Caius College* by John Venn, 1989, p. 272.

page 64. *it was abolished*...Cambridge abolished the "ten-year man" regulation in 1858. *Bedders, Bulldogs and Bedells: A Cambridge Glossary* by Frank H. Stubbings. Cambridge University Press, 1995.

page 65. *British Association*...The British Association for the Advancement of Science was founded in 1831 to promote science, which at the time was suffering a decline in popularity and interest. The BA, as it was known, quickly established a solid reputation for itself. It is now known as the British Science Association.

page 65. *Smithsonian Institution in 1850*...Cox , 2002, p. 56.

page 67. *American Storm Controversy*...Thomas, 1991, p.102.

page 67. *should be organized*...Monmonier, 1999, p. 39.

page 67. *east to west*...Cox, 2002, p. 56.

page 67. *and other newspapers*...Monmonier, 1999.

page 67. *groundwork in place*...The telegraph system was actually the smaller of two programs organized by Henry. In order to get a grasp on the real nature of weather systems, he needed a great many observations at the same time and over a wide area—far more than his telegraph operators could provide. And so the other, larger program was another network of weather observers—their numbers varied between two and five hundred—who mailed detailed monthly weather reports to Washington. This network gathered huge amounts of useful data.

page 68. *the Best Lifeboat?*...*Nautical Magazine*, Vol. 20, 1851, p. 660.

page 69. *lifeboat was born*...Encyclopaedia Brittanica, Vol. 14, 1888, p. 572 (available on Google Books); see also *Wikipedia*, the reference to James Beeching under "Lifeboat (rescue)."

page 70. *testimonial for Saxby*...Testimonial from William Fenwick with Saxby's application to the navy. Fenwick took over command of the *Akbar* in December 1847. *Irish Quarterly Review*, Vol. 7, March 1857.

page 70. *in the North Sea*...*An Address to the Ship Owners of Liverpool*. Saxby also mentions two other sea trips, one up the Thames to London Bridge, and a trip by steamer from Chepstow to Bristol.

Chapter 3: Storm Warnings

page 74. *his personal finances*...During his governorship of New Zealand and as captain of HMS *Arrogant*, FitzRoy had been expected to entertain on a lavish scale, one he could barely afford. His salary at the Met Office was not generous—a mere £600, although it was raised to £800 in 1863.

page 75. *it was not to last*...Gribbin, 2003, p. 240.

page 75. *not incompatible*...Cox, 2002, p. 39.

page 75. *between Henry and Matthew Maury*...Cox, 2002, p. 53.

page 76. *on his territory*...Both claimed first dibs on studying the Gulf Stream.

page 77. *conference in Brussels*...For a full account of the lead-up to the Brussels Conference, see Williams, Frances Leigh, *Matthew Fontaine Maury: Scientist of the Sea*. New Brunswick & New Jersey: Rutgers University Press, 1963.

page 77. *their weather observations*...Ibid.

page 77–78. *laughter from the backbenches*...Gribbin, 2003, p. 252.

page 78. *around the coast each year*...Burton, 1986, p. 158.

page 79. *the last day of the meeting*...Friday, September 22.

page 80. *using iron plates*...Scoresby's criticism of the compasses was met with great alarm by shipowners, who thought he was sounding the death knell for iron ships and implying that a return to wooden hulls was the answer. He was implying no such thing, and had to make a concerted effort put their minds at rest.

page 80. *a bitter disappointment*...The *Report of the BA* for 1854 includes, on p. 49, an article by William Scoresby, "On the Loss of the *Tayleur*, and the Changes in the

Action of Compasses in Iron Ships." It also includes, under *Notices and Abstracts of Miscellaneous Communications to the Sections*, on p. 161: "On Mechanical Appliances on Board Merchant Ships. By Mr. Saxby." The report includes only the title—the article itself was not published. In *An Address to the Ship Owners of Liverpool*, Saxby explains his reasons for not reading it at the BA meeting.

page 81. *One of the outcomes of the BA meeting*...A few months after the BA meeting, an undaunted Saxby self-published his paper as a pamphlet titled *An Address to the Ship Owners of Liverpool*. In addition to the compass problem and the inadequacies of current methods of correcting compasses, he discussed a number of other safety issues affecting merchant shipping. He stressed the importance of people's wearing lifebuoys during emergencies and the need for fire "annihilators," and called for a better way of estimating ships' tonnages and for a better design of their rudders—the *Tayleur* had been very slow to wear around due to an inadequate rudder. But the pamphlet was also, in part, a promotional tool for some of his inventions, which included a rapid method of reducing topsails that he had patented. He had also made improvements to the design of bitts—the stout oak and iron posts to which a ship's anchor cable is secured—and of windlasses to make it easier to veer chain and reduce the strain on ground tackle, citing disasters with these devices he had witnessed earlier in life. Another of his patents was for an improved method for lowering ships' boats and for freeing them once they had been lowered, again based on what he had learned from recent accidents that had occurred during emergency operations. Saxby was confident that he had the backing of Liverpool's shipowners, who had already accepted a number of his inventions with what he describes as "unqualified approval."

page 82. *totally impractical*...*Nautical Magazine*, Vol. 27, 1858, pp. 153–55.

page 82. *ample crinoline*..."Lady Passengers and the Ship's Compass." *Nautical Magazine*, 27 (1858), p. 275.

page 83. *peppered his prose*...Despite the many criticisms of Maury, he deserves much credit as a practical oceanographer. His *Wind and Current Charts* in particular were a real breakthrough. FitzRoy wrote of him: "Although, as I am aware, he occasionally theorises when he has not facts enough for philosophy, as a practical man he has been guided by plain principles, intelligible to seamen generally."

page 84. *Medicine and Surgery*...Ferrel, William. "An Essay on the Winds and Currents of the Ocean." *Nashville Journal of Medicine and Surgery*, Vol. 11, 1856, pp. 287–301, and 375–89. It's perhaps not as surprising a venue for an article on weather as it sounds: at the time there was a common belief that the weather had serious impacts on human health.

page 84. *to the left in the southern*...Ferrel, William. "The Influence of the Earth's Rotation Upon the Relative Motion of Bodies Near its Surface." *Astronomical Journal*, Vol. 5, 1868, p. 99.

page 85. *the centre of a storm*...There is one other element at work too: centrifugal force, which affects any rotating body and reinforces the Coriolis effect by deflecting a rotating wind outwards from the centre of a storm.

page 86. *autocratic rule*...In the early 1840s Arago had investigated the idea that the moon and other celestial bodies influence weather, and dismissed any such effects as being "almost insensible." Cox, 2002, p. 87.

page 86. *as far away as Australia*...Rowland, 1970, p. 64.

page 86. *to operate and maintain them*…Until now, the Lords of the Admiralty were also asking themselves a more basic question: did Britain really need a steam-powered battle fleet? None of the country's traditional foes was building one at the time, so why scrap an existing sailing fleet that had cost a fortune to build if there was no foreseeable need?

page 87. *on these ships were better*…Dickinson, 2007.

page 87. *replaced by the Portsmouth College*…Dickinson, 2007.

page 88. *difference in status*…Dickinson, 2007.

page 88. *from 440 to 862*…Walton, 2004, p. 4.

page 89. *new machinery*…The Royal Naval Engineering College at Devonport was not established until 1880.

page 89. *Russian POWs*…National Archives, ADM 101/256. The regulation also made instructors out of a class of men who were to a large extent ill-equipped for preparing young men to sail, navigate, and operate ships that were becoming ever more technical. Many of them barely knew one end of a sextant from the other. Some had never seen a chart! And even with the ranks of the instructors filled with clergy, in the two decades that followed the 1837 closure of the naval college there were simply not enough naval instructors to go round. The situation became so bad that many midshipmen reached their early- to mid-twenties so ignorant of navigation that they had little hope of ever passing their lieutenant's exam.

page 89. *HMS* Britannia…which eventually gave its name to the naval college at Dartmouth, England, that exists today.

page 90. *In his application*…Almost everything that follows about Saxby's career with the Royal Navy, in this and later chapters, can be found at the British National Archives, Kew, London. See ADM196/68, ADM1/ 5707, ADM1/5960, ADM1/6138, ADM1/6102, ADM1/6090, ADM1/6122, and ADM1/6132.

page 92. *living with their parents*…In 1858, Saxby's children were aged 14, 17, 20, 22, and 27.

page 93. *the navy's new steam engines*…Between 1832 and 1869, twenty-six vessels were built in the Sheerness yard—mostly steamers, but a handful of sailing vessels, too. Sheerness Dockyard remained in operation until March 1960, when the land was sold into private hands. It is now operated as a commercial port.

page 94. *leaders were hanged*…Gill, C. *The Naval Mutinies of 1797*. Manchester: Manchester University Press, 1913.

page 94. *the last place God made*…Fremantle, Admiral Sir Edmund R. *The Navy as I Have Known It, 1849–1899*. London: Cassell & Co. 1904.

page 95. *the gold rush in Australia*…*Royal Charter*, 2,719 tons, was built in 1855 by Gibbs, Bright and Company on the River Dee. Charles Dickens wrote an account of the disaster's aftermath in *The Uncommercial Traveller*.

page 95. *middle of the town*…*Times*, Thursday, Oct 27, 1859.

page 96. *a storm warning system*…Anderson, 2005, p. 120.

page 97. *Nairn in Scotland*…*Times*, Feb. 7, 1861.

page 97. *number of shipwrecks*…Burton, 1986, p. 161.

page 97. *the architect*…ODNB.

page 97. *Responsibility*…February 1861 letter from FitzRoy to T. H. Farrer. Farrer was appointed an assistant secretary to the Board of Trade in 1853. National Archives MT9/13/2883.

Chapter 4: Barking at the Moon

page 101. *of the one object...Saxby's Weather System*, p. 41.

page 104. *knows what he is about*...From 1860 to 1861, Saxby publicized his ideas in a long series of articles in the *Nautical Magazine* titled "Lunar Equinoctials".

page 104. *in relation to the earth*...Anderson, 2005, p. 47.

page 104. *with unerring accuracy*...Journal of Agriculture, 1863, pp. 390–94.

page 105. *"The Coming Winter and the Weather"*...*Nautical Magazine*, Vol. 31, December 1862.

page 106. *crying wolf*...In December 1863 a rumour spread in Portugal and Gibraltar that FitzRoy had predicted an extraordinary storm. The superstitious and the devout flocked to the Roman Catholic church in Gibraltar, offering up prayers in the hope of averting the storm. In Lisbon, fishermen were so alarmed at the prospect that they rushed back to port, leaving valuable fishing gear at sea. The rumour was false, and throughout the predicted period the weather was unusually fine, but the identity of its originator was never discovered. FitzRoy was incensed that his name had been attached to what he called "pseudo-prophecies of tempests," and "absurd but injurious predictions." He had his suspicions about who was responsible but never pointed the finger at anyone. Saxby's name has been put forward as a possible culprit, but this seems highly unlikely. Although Saxby's prophecies often went unfulfilled, it's unthinkable that he would have embellished them by claiming they had the support of FitzRoy. (FitzRoy, letters to the Times, January 1 and 18, 1864.)

page 107. *influenced by the moon*...*Saxby's Weather System*, p. 56.

page 108. *my further investigations*...In *Saxby's Weather System*, p. 14, he mentions a meeting with FitzRoy at the latter's office in February 1861.

page 108. *my apparent presumption*...*Saxby's Weather System*, pp. 13–14.

page 108. *another on marine steam engines*...Saxby's *The Projection and Calculation of the Sphere* appeared in 1861 and was followed a year later by *The Study of Steam and the Marine Engine*.

page 110. *continued to publish them*...Gribbin, 2003, p. 263.

page 110. *not be issuing forecasts*...Ibid., p. 278.

page 112. *Lunar Influence on Weather*...Saxby retitled his book because a Dutch author, F. H. Klein, published a pamphlet in 1863, translated into English as *The Foretelling of the Weather in Connexion with Meteorological Observations*.

page 112. *the barometer was falling rapidly*...*Saxby's Weather System*, pp. 4–5.

page 114. *for the said truth*...Ibid., p. 19.

page 114. *heaping scorn on their ideas*...In 1864 Saxby also devoted time to writing about the study of the sea in a very lengthy and somewhat whimsical series of articles that discuss the pleasures to be enjoyed at the coast, and his thoughts on geology. His purpose was to encourage those with enquiring minds to study the coast of Britain. He even invented a new science, which he called Undavorology from Unda—wave and Voro—eat into, devour—about the destructive action of waves on the shore. He proposed that the Scilly Isles were separated from England by water action rather than some "great volcanic disturbance." The new science did not catch on, and, like other ideas that poor old Saxby hoped to popularize, disappeared into the dustbin of history.

page 115. *the West Indies…Saxby's Weather System.* p. 83.

page 115. *sky of cloud…*Sir John Herschel, "The Weather and Weather Prophets," Good Words, Vol. 5, 1864, pp. 57–64.

page 116. *his programs…*Fleming, 1990, p. 148.

page 117. *a quarter-deck…*Gribbin, 2003, p. 5.

page 118. *headway for decades…*Burton, 1986, p. 175.

page 119. *have been deduced…*Quoted in Fleming, 1990, p. 148.

page 119. *fully vindicated…*In the UK, FitzRoy finally became a household name in February 2002, when the UK Meteorological Office renamed one of the sea areas featured in its shipping forecasts in his honour. Finisterre became known as FitzRoy, and now his name can be heard four times a day on BBC Radio 4 as the forecaster reels off the litany of gale warnings and other delights that await seafarers around the British Isles.

page 119. *with his performance…*Saxby was "officially" stationed on *Cressy* (21/5/58–24/4/59), *Minotaur* (25/4/59–30/9/59), *Cumberland* (1/10/59–31/3/63, 1/4/63–31/12/68), and *Agincourt* (1/1/69–9/6/69). ADM 196/68.

page 119. *a glowing report…*January 15, 1860.

page 119. *encroaching deafness…Saxby's Weather System*, p. 15.

page 120. *the engineer officers…*In November 1863, an incident at Sheerness involving an engineer officer caught the attention of the British public. The Royal Navy had decided to sell off one of its steam-powered sloops, the *Victor*, which had never performed well. The vessel was stripped of its armaments, masts, and rigging, then refitted and sold to a London firm. But Matthew Fontaine Maury, who had been dispatched to England by the Confederacy, secretly purchased the *Victor* for service as a commerce raider. The British authorities somehow got wind of this deal, and the new owners in turn realized their plan had been exposed. Under cover of darkness, the ship weighed anchor at Sheerness and steamed down the Thames estuary. Once out at sea it was boarded by Confederate naval officers, who hoisted the Confederate flag and renamed the ship CSS *Rappahannock*. It got no further than Calais, however, where it was detained by the French Government for the duration of the Civil War.

A high-ranking engineering colleague of Saxby's, William Rumble, an Inspector of Machinery Afloat, was implicated in the *Rappahannock* affair. One of Rumble's duties—which would have brought him into close contact with Saxby—was the examination of Engineer Students. Rumble had supervised much of the ship's refit, and on the very evening the vessel had slipped quietly out of the Thames he had been on board until five o' clock. Rumble was brought to trial under the *Foreign Enlistment Act*—Britain was officially neutral in the American conflict—but was acquitted (much to the chagrin of the London *Times*), although he was never employed by the navy again. Unfortunately, there is no record of what Saxby knew or thought of this affair. National Archives, FO5/1052, ADM7/627.

page 122. *the wide world…London Standard*, Dec. 25, 1868.

page 122. *moon at 2 p.m.…*quoted in Heidorn, 2010.

page 123. *office be abolished…*E. J. Reed to Sir Spencer Robinson, April 5, 1869.

page 124. *Cambridge University…*The university was a parliamentary constituency until 1950.

page 125. *a local newspaper…*the *Halifax Evening Express and Commercial Record*.

page 127. *a rather curious hurricane*...Ludlum, 1963, p 103.

page 127. *reaching one hundred knots*...Murnane and Liu, 2004, p. 203.

page 127. *tore out the gables*...A gabled roof is more vulnerable to strong winds than a hip roof. If the flat surface of a gable is perpendicular to the wind it offers maximum resistance, whereas the sloping surfaces of a hip roof will deflect the wind from any direction.

page 128. *workable weather network*...Abbe, from the *Dictionary of American Biography*, and from Cox, 2002. Abbe's experience placed him in a unique position that made him the perfect choice within a few years to direct the Weather Service of the United States, beginning on Jan. 3, 1871. In the U.S. the term "forecast" was not introduced until 1889.

page 128. *"Old Probabilities"*...When Mark Twain concocted a spoof forecast of the notoriously fickle weather of New England in 1876, he was poking gentle fun at Cleveland Abbe: *Probably northeast to southwest winds, varying to the southward and westward and eastward and points between, high and low barometer swapping around from place to place, probable areas of rain, snow, hail, and drought, succeeded or preceded by earthquakes, with thunder and lightning...But it is possible that the programme may be wholly changed in the meantime.*

Chapter 5: Deluge

The contemporary accounts in this chapter are drawn from the columns of the *New York Times* of October, 1869. Additional shipping details are from the *New York Maritime Register* of that month.

page 137. *something like this*...This scenario of the storm's likely origin was told me by Chris Fogarty of the Canadian Hurricane Centre. He stresses that it is only a theory, but it is based on the best evidence available.

page 139. *the heat engine*...Sheets and Williams, 2001, p. 33.

page 147. *gales and thunderstorms*...Nearly all these transitioning storms remain serious storms. They may put on a burst of speed, travelling as much as two, three, or even five times faster than when they were in the tropics. This, of course, increases the effective wind speed on the right of the storm even more.

page 147. *a cold front*...No one knew this at the time. The storm was later reconstructed by Abraham et al., in 1998. Ludlum (1963) gave an earlier account of the meteorology of the storm.

page 151. *at Fall River*...Ludlum, 1963.

page 153. *the bodies recovered*...Eventually, construction of the Hoosac Tunnel cost the lives of 195 men.

Chapter 6: Landfall

The accounts of the storm in this and the following chapter are drawn mostly from contemporary reports in the following newspapers: *Saint John Morning Freeman, Saint John Daily Telegraph & Morning Journal, St. Croix Courier, Daily Morning News* (St. John), *Morning Chronicle, Commercial and Shipping Gazette* (Quebec), *New Brunswick Reporter & Fredericton Advertiser*, the *NovaScotian, Morning Chronicle* (Halifax), *Halifax Daily Reporter and Times,* and the *Halifax Evening Express and Commercial Record.*

page 158. *the paddlewheeler* New York...This steamer was built in 1852 at Clayton, New York, on the St. Lawrence River and operated on the Great Lakes before coming to Maine in 1860. In the Civil War it served as a transport and prisoner exchange boat, then returned to Maine. It finally burned in the 1890s in Philadelphia.

page 165. *121 vessels were beached*...From the *Morning Chronicle*, October 14, 1869, "Shipping Disasters by the Storm."

page 170. *a brand-new vessel*...On the day of its launching on September 9 there was an unfortunate event that may have cast *Genii* as an unlucky ship among superstitious seafarers. During its launching the barque fell over on its side. On the following tide it righted itself and all appeared to be well. A number of sailors refused to sail on the ship, however, saying that it looked as though it wouldn't sail well in heavy weather. This may explain why it sailed from St. Andrews with so few men. Additional crew would have been needed to sail it across the Atlantic, and they must have been intended to join the ship at New River. The best and fullest account of this tragedy is by Medcof, 1966.

page 173. *eleven men lost their lives*...Seven of the bodies of *Genii*'s crew, including that of Captain Bailey, were found on the shore on Tuesday morning. Two more were found on Wednesday, one of them George McVicker, whose body was so badly bruised that it was almost unrecognizable. The final one was washed ashore ten days after the storm. (Medcof, 1966.)

page 175. *piles higher and higher*...Sheets and Williams, 2001, p. 73.

Chapter 7: Storm Surge

page 182. *drives the tides*...There are atmospheric tides too, but they are caused mainly by the sun, and scientists generally agree they do not affect weather significantly.

page 182–83. *anywhere else in the world*...The Bay of Fundy can claim its record by only a narrow margin, however. The tides at Ungava Bay are just a few centimetres lower (but see note for Chapter 9).

page 184. *the bay's dimensions*...C.O'Reilly, personal communication.

page 184. *to ever-greater heights*...For the same reasons, other bays with similar shape and dimensions to the Bay of Fundy—Ungava Bay in northern Quebec, the Severn Estuary/Bristol Channel in England and the Gulf of St. Malo in France, for instance—also have extreme tidal ranges, although they can't match the giant tides of Fundy. There are other places, too: King Sound in Western Australia, and the Gulf of Khambhat in India.

page 191. *in new locations*...I have been unable to find the source of this story. None of the published accounts provide the names of the husband and wife or the location.

page 196. *no reports of damage*...Many years later Mrs. MacCallum Grant, whose father owned the Beechwood estate, recalled that the Saxby gale was one of her earliest memories: "...even after all these years I can remember the uncanny howling and screeching of the wind and the torrents of rain." Nova Scotia Archives and Records Management (NSARM): Misc-Storms-Saxby Gale.

Chapter 8: Aftermath

page 199. *both totally gutted*...Whalen, 1995.

page 199. *high tide on Monday night*...Saint John *Daily Telegraph and Morning Journal*, October 12, 1869. Storm surge can be both positive, causing excessively high tides, or negative, causing excessively low tides like this one. Such low tides can be of source of grief to mariners, but are of little concern to land-dwellers.

page 202. *a fire hazard for many years*...Hutchinson, 1912.

page 205. *victims were drowned*...Whalen, 1995.

page 205. Punch *magazine*...October 16, 1869.

page 209. *Saxby in effigy*...Reading Mercury, November 27, 1869.

page 210. *substantially fulfilled*...Daily Morning News, Thursday October 14.

page 211. *letter to the editor*...Nautical Magazine, Vol. 38, 1869, p. 338.

page 213. *spent his last years*...Of Stephen and Ann Saxby's six children, Stephen Henry and Gavin Frank both joined the church, Elizabeth Mary Ann became an author, Henry Linckmyer a physician in Scotland, and Gordon Harding a soldier in New Zealand. There is no record of what career Mary Louisa pursued. Henry Linckmyer Saxby was also a keen ornithologist who contributed a number of papers to the journal *Zoologist*. He died at the early age of 37. One year later, in 1874 his older brother Stephen Henry published Henry Linckmyer's book, *The Birds of Shetland*. Stephen Henry, like his father, entered Gonville and Caius College Cambridge, and graduated in 1854. He had a keen interest in astronomy, perhaps inspired by his father's interest. He wrote several papers for the Royal Astronomical Society, and a year before his death in 1886 he was elected a fellow of that society.

page 213. *Greenwich in 1873*...Clowes, Sir W. Laird. *The Royal Navy: A History*. Vol. 7. London: Sampson Low, Marston & Co., 1903. p. 71.

page 214. *the U.S. Weather Bureau*...There had been various attempts to produce weather maps in the early part of the century, but by the 1870s the first maps using isobars, similar to those with which we are familiar today, appeared. (Thomas, 1991, p. 79).

page 216. *from the West Indies*...A telegraph cable had been laid between Havana and Key West in 1867, but at the time no one in Cuba was able to provide hurricane warnings.

page 217. *city of Galveston*...Larson, p. 102.

Chapter 9: Why Saxby Still Matters

page 219. *about to be repeated?*...Emily Bowers, "Storm Surge," Halifax *Chronicle Herald*, October 5, 2002.

page 220. *a similar storm surge*...Abraham et al, 1998.

page 220. *dyke that protected it*...Desplanque and Mossman, 1999; Denis, A. S. "The Tropical Cyclone in the Maritimes." B.Sc. thesis, Acadia University, May 1949.

page 221. *taking some six hundred lives*...Ruffman, April 1999.

page 223. *on the scale of Saxby's storm*...Parkes et al., 1997.

page 224. *Brooklyn, New York*...C. Fogarty, personal communication.

page 224. *details of Stephen Saxby*...Letters from Cleveland Abbe (Weather bureau, Washington), and letters from the Met. Office, London. William Francis Ganong Fonds, S217, F1, Item 1, New Brunswick Museum.

page 224. *the first "scientific" account*... Hutchinson, 1912.

page 225. *as superintendent*...William Bell Dawson came from a family of talented scientists. His father, Sir John William Dawson, and his brother George Mercer Dawson were both eminent geologists.

page 226. *makes them unique*...Parkes et al., 1997.

page 227. *anywhere else in the world*...This is the maximum predicted range, plus or minus 0.4 metres, (1.3 feet) in the Minas Basin for the period 1998 to 2016. See: O'Reilly, C. T.; Solvason, R.; Solomon, C., 2005. "Where are the World's Largest Tides?" in *BIO: 2004 in Review*. J. Ryan (Ed.), pp. 44–46. For those interested in records, Ungava Bay in Quebec ties with the Bay of Fundy for *extreme* high tides, but the Bay of Fundy *regularly* has the highest tides (C. O'Reilly, personal communication).

page 228. *to 90 knots (167 km/h)*...Murnane & Liu, 2004, table 7.5, page 188.

page 231. *two feet (0.6 metres) higher*...Parkes et al., 1999.

page 235. *in September 2014*...There is another interaction between the motions of sun, moon, and earth that may influence exceptionally high tides. In the late 1990s, a hydraulic engineer who was an expert on storm tides in the Bay of Fundy, the late Con Desplanque, along with David Mossman, a geoscientist at Mount Allison University, New Brunswick, described another long-term rhythm of the tides. Superimposed on the regular rhythms already described here, there is another cycle, known to the Babylonians as far back as 800 BCE (who used it to predict eclipses) and well-known to astronomers today. It is known as the *Saros* cycle. Precisely every eighteen years, eleven days, and eight hours (18.03 years), the sun, earth, and moon return to exactly the same perfect alignment, providing not only the necessary conditions for eclipses, but also ideal conditions for generating large tides over a period of weeks or months. A *Saros* occurred in 1958, and backtracking five *Saros* cycles from that reference point takes us to 1869, and the Saxby tide.

Working back another six *Saros* cycles takes us to another devastating storm tide on November 3 to 4, 1759. Desplanque and Mossman have found other matches: an extremely high tide in Moncton on October 12, 1887; the Groundhog Day Gale of February 2, 1976; and storm tides between December 20 and 22, 1995, all coincided with a *Saros*. There is a *Saros* peak in 2012–2013. (Desplanque and Mossman, 1999 and 2004.)

The Canadian Hydrographic Service, however, uses the 18.6 year cycle when compiling its tidal predictions, not the *Saros* cycle. (C. O'Reilly, personal communication).

page 243. *put EMO's plans to the test*...For details of the 2003 storm see www. novaweather.net/Flood_2003.html. This storm took the lives of two people when their car plunged into the LaHave River near Bridgewater, Nova Scotia. It caused disastrous flooding in all four Atlantic provinces, and was closely followed by two other events that also put EMO's plans to the test: Hurricane Juan in September 2003 and White Juan in February 2004.

Glossary

Atlantic Multi-decadal Oscillation (AMO)—A cycle of alternating warm and cool sea-surface temperatures in the North Atlantic Ocean that repeats every twenty years or so and may influence the frequency of severe hurricanes.

Apogee—The part of the moon's orbit when it is farthest from the earth.

Astronomical tide—The rise and fall of the tides that would result (in theory) from the gravitational pull of the sun and moon alone, if the atmosphere did not have any influence.

Coriolis effect—An apparent force, caused by the earth's rotation, that deflects moving objects (and winds) to the right in the northern hemisphere, and to the left in the southern. It gives tropical storms their spinning motion.

El Niño—A warming of the equatorial waters of the eastern Pacific Ocean that usually happens every three to seven years. It generates strong winds in the upper atmosphere that can hinder the development of tropical cyclones in the Atlantic.

Frontal storm—A storm formed when warm and cold air masses meet and form "fronts." The winds blow counter-clockwise around a centre of low atmospheric pressure.

Hurricane—A large, rotating storm in the Atlantic Ocean (and also in the central and eastern Pacific Ocean), north of the equator, originating in the tropics, with maximum sustained surface wind speeds of 64 knots (118 km/h) or more.

Jet stream—A narrow, fast-moving westerly band of air in the upper atmosphere.

Knot—A unit of speed of one nautical mile (two thousand yards) per hour. Equivalent to about 1.14 mph or 1.8 km/h.

La Niña—A cooling of the equatorial waters of the eastern Pacific Ocean that happens between El Niño events. It weakens winds in the upper atmosphere, which favours the development of tropical cyclones in the Atlantic.

Latent heat—The heat that is absorbed by water (or other substances) when it becomes a vapour, and is released when the vapour condenses back into water droplets. It provides the fuel that drives a hurricane.

Neap tides—These occur near the moon's first and last quarter, when the gravitational pull of the moon and sun work at right angles to each other, and the range of the tides is at its smallest.

Perigee—The part of the moon's orbit when it is nearest the earth.

Post-tropical transition—The change a hurricane undergoes as it moves northward and loses its tropical character.

Spring tides—These occur near full and new moon, when the gravitational pull of the moon and sun work in conjunction to generate the highest high tides and the lowest low tides.

Storm surge—A rise (or fall) in sea level along a coast, caused by strong winds and changes in atmospheric pressure during a storm.

Tidal range—The difference in height between successive high and low tides.

Tropical cyclone—Any kind of tropical rotating low pressure weather system. Winds blow around it counterclockwise in the northern hemisphere.

Tropical depression—A tropical cyclone with maximum sustained surface winds of 34 knots (63 km/h) or less.

Tropical storm—A tropical cyclone with maximum sustained surface winds between 34 and 63 knots (63 and 118 km/h).

Tropical wave—A kink or bend in the easterly trade winds of the tropical Atlantic that often develops into a cyclone. Also known as an "easterly wave."

Trough—An elongated area of low atmospheric pressure.

Typhoon—The name given to a tropical cyclone over the western Pacific Ocean, north of the equator.

Wind shear—A sudden change of wind speed or direction with altitude.

Bibliography

Books

Anderson, Katharine. *Predicting the Weather: Victorians and the Science of Meteorology*. Chicago and London: University of Chicago Press, 2005.

Bleakney, J. Sherman. *Sods, Soil, and Spades: The Acadians at Grand Pré and Their Dykeland Legacy*. Montreal and Kingston: McGill-Queen's University Press, 2004.

Browne, Walter Lord. *The Moon and the Weather: The Probability of Lunar Influence Reconsidered*. 2nd Edition. London: Baillière, Tindall, and Cox, 1885.

Cox, John D. *Storm Watchers: The Turbulent History of Weather Prediction from Franklin's Kite to El Niño*. Hoboken, N.J.: John Wiley & Sons, 2002.

Dawson, W. Bell. *Tide Levels and Datum Planes in Eastern Canada: from determinations by the Tidal Current Survey up to the year 1917*. Ottawa: Department of the Naval Service, 1917.

Dickinson, H. W. *Educating the Royal Navy: Eighteenth- and Nineteenth-Century Education for Officers*. London and New York: Routledge, 2007.

Dictionary of American Biography. 1928.

Dictionary of Scientific Biography. 1970.

Elsner, James B., and A. Birol Kara. *Hurricanes of the North Atlantic: Climate and Society*. New York: Oxford University Press, 1999.

FitzRoy, Robert. *The Weather Book: A Manual of Practical Meteorology*. London: Longman, Green, Longman, Roberts, & Green, 1863.

Fleming, James Rodger. *Meteorology in America 1800–1870*. Baltimore and London: The Johns Hopkins University Press, 1990.

Gribbin, John and Mary. *FitzRoy: The Remarkable Story of Darwin's Captain and the Invention of the Weather Forecast*. London: Review/Headline Book Publishing, 2003.

Hearn, Chester G. *Tracks in the Sea: Matthew Fontaine Maury and the Mapping of the Oceans*. Camden, Maine: International Marine, 2002.

Kennedy, John. *The History of Steam Navigation*. Liverpool: Charles Birchall, 1903.

Larson, Erik. *Isaac's Storm: A Man, a Time, and the Deadliest Hurricane in History*. New York: Vintage Books, 2000.

Ludlum, D. M. *Early American Hurricanes, 1492–1870*. Boston: American Meteorological Society, 1963.

Maury, Matthew Fontaine. *The Physical Geography of the Sea and its Meteorology*. 1855. New York: Harper & Brothers, 1861.

Monmonier, Mark. *Air Apparent: How Meteorologists Learned to Map, Predict, and Dramatize Weather.* Chicago and London: University of Chicago Press, 1999.

Mooney, Chris. *Storm World: Hurricanes, Politics, and the Battle over Global Warming.* Orlando: Harcourt, 2007.

Murnane, Richard J., and Kam-Biu Liu, eds. *Hurricanes and Typhoons: Past, Present, and Future.* New York: Columbia University Press, 2004.

Oxford Dictionary of National Biography. 2004.

Penn, Geoffrey. *"Up Funnel, Down Screw!": The Story of the Naval Engineer.* London: Hollis & Carter, 1955.

Pearce, Alfred J. *The Weather Guide-Book: A Concise Exposition of Astronomic-Meteorology.* London: Simpkin, Marshall & Co., 1864.

Ring, Ken. *Predicting the Weather by the Moon.* Glastonbury: Gothic Image Publications, 2002.

Rowland, K. T. *Steam at Sea: A History of Steam Navigation.* Newton Abbot: David & Charles, 1970.

Saxby, S. M. *Saxby's Weather System: or Lunar Influence on Weather.* 2nd ed. London: Longman, Green, Longman, Roberts, & Green, 1864.

——. *The Study of Steam and the Marine Engine for Young Sea Officers in H.M. Navy the Merchant Navy, Etc. being a Complete Initiation into a Knowledge of Principles and their Application to Practice.* London: Longman, Green, Longman, and Roberts, 1862. (Elibron Classics Replica Edition, 2006.)

Sheets, Bob, and Jack Williams. *Hurricane Watch: Forecasting the Deadliest Storms on Earth.* New York: Vintage, 2001.

Thomas, Morley K. *The Beginnings of Canadian Meteorology.* Toronto: ECW Press, 1991.

Articles

Abbe, Cleveland. "A Chronological Outline of the History of Meteorology in the United States of North America." *Monthly Weather Review* 37.3 (1909): 87–89; 37.4 (1909): 146–49; 37.5 (1909): 178–80.

Abraham, Jim, George Parkes, and Peter Bowyer. "The Transition of the 'Saxby Gale' into an Extratropical Storm." *The 23rd Conference on Hurricanes and Tropical Meteorology.* Dallas, Texas: Amer. Meteor. Soc., 1998. 795–98.

Allison, Frederick. "Nova Scotian Meteorology." *Proc. Trans. NS Inst. Sci.* 4.3 (1876–77): 300.

Anon. "Saxby's Weather System." *The Journal of Agriculture, July 1861–March 1863.* Edinburgh and London: William Blackwood and Sons, 1863. 390–94.

Burton, Jim. "Robert FitzRoy and the Early History of the Meteorological Office." *British Journal for the History of Science* 19 (1986): 147–76.

Desplanque, Con, and David J. Mossman. "Storm Tides of the Fundy." *Geographical Review* 89.1 (January 1999): 23–33.

——. "Tides and their seminal impact on the geology, geography, history, and socio-economics of the Bay of Fundy, eastern Canada." *Atlantic Geology* 40.1 (2004): 1–130.

FitzRoy, Rear-Admiral. "Remarks on the late Storms of October 25–26 and November 1, 1859." *Proc. Roy. Soc.* 10 (1860): 222–24.

FitzRoy, Robert. "On British Storms." *Report of the British Association for the Advancement of Science.* Oxford, June/July 1860: 39–44.

Ganong, W. F. "A Preliminary Study of the Saxby Gale." *Bulletin of the Natural History Society of New Brunswick* 29.6 (1911): 325–30.

Graney, Christopher M. "Coriolis Effect, Two Centuries Before Coriolis." *Physics Today* 64.8 (August 2011): 8. Available online at http://dx.doi.org/10.1063/PT.3.1195.

Heidorn, Keith C. "The Saxby Gale: A Lucky Guess?" *The Weather Doctor*, October 2010. www.islandnet.com/~see/weather/almanac/arc2010/alm10oct.htm.

Hughes, Patrick. "FitzRoy the Forecaster: Prophet without Honor." *Weatherwise* 41.4 (August 1988): 200–204.

Hutchinson, D. L. "The Saxby Gale." *Transactions of the Canadian Institute* 9 (1912): 253–59.

Ludlum, David. "The Espy-Redfield Dispute." *Weatherwise* 22 (December 1969): 224–29, 245, 261.

Medcof, J. C. "Loss of the Barque *Genii* at New River, N.B. During the Saxby Gale, 1869." *Collections of the New Brunswick Historical Society* 19 (1966): 17. Available online at www.rootsweb.ancestry.com/~nbpstgeo/stge5genii.htm.

Olmstead, Denison. "A Biographical Memoir of William C. Redfield." *American Journal of Science and Arts* 2.24 (1857): 355–73.

Parkes, G. S., L. A. Ketch, and C. T. O'Reilly. "Storm Surge Events in the Maritimes." *Proceedings of the 1997 Canadian Coastal Conference.* Revised June 1998.

Parkes, George S., Lorne A. Ketch, Charles T. O'Reilly, John Shaw, and Alan Ruffman. "The Saxby Gale of 1869 in the Canadian Maritimes: a case study of flooding potential in the Bay of Fundy." *The 23rd Conference on Hurricanes and Tropical Meteorology.* Dallas, Texas: Amer. Meteor. Soc., 1999.

Rao, Joe. "Moonstruck Meteorology." *Weatherwise* 55.5 (Sept./Oct. 2002): 23–29.

Redfield, William C. "Remarks on the Prevailing Storms of the Atlantic Coast of the North American States." *American Journal of Science and Arts* 1.20 (1831): 17–51.

Ruffman, Alan. "The Saxby Gale: an October 4–5, 1869 Tropical Cyclone with a Hybrid Twist." *Argonauta, the Newsletter of the Canadian Nautical Research Society* 16.2 (April 1999): 3–5.

———. "A Multi-disciplinary and Inter-Scientific Study of the Saxby Gale: an October 4–5, 1869 Hybrid Hurricane and Record Storm Surge." *CMOS Bulletin* 27.3 (June 1999): 67–73. Available online at www.shunpiking.com/ol0103/03SaxbyGale11869RUFFMAN.htm.

Saxby, S. M. "Which is the Best Lifeboat?" *Nautical Magazine* 20 (1851): 660.

———. "Terrestrial Magnetism and the Law of Storms." *Nautical Magazine* 21 (1852): 152.

———. "On Lowering Boats." *Nautical Magazine* 22 (1853): 443.

———. *An address to the ship owners of Liverpool.* 1854. Self-published pamphlet.

———. "Saxby's Stopper." *Nautical Magazine* 24 (1855): 92.

———. "Saxby's Spherograph." *Nautical Magazine* 26 (1857): 198.

———. "Suggestions for Avoiding Collisions at Sea." *Nautical Magazine* 27 (1858): 153–55.

———. "Lunar Equinoctials Affecting the Weather." *Nautical Magazine* 29 (1860): 138, 355, 390, 482, 609.

———. "Lunar Equinoctials." *Nautical Magazine* 30 (1861): 69, 164, 190, 263, 342, 384.

———. "Cyclones and Saxby's Weather System." *Nautical Magazine* 31 (1862): 32.

———. "Saxby's Weather System." *Nautical Magazine* 31 (1862): 364.

———. "The Coming Winter and the Weather." *Nautical Magazine* 31 (1862): 664–71.

———. "Weather Warnings and a Great Day Auroral Storm." *Nautical Magazine* 32 (1863): 146–55.

———. "Coals Used in Steamers." *Nautical Magazine* 38 (1869): 133, 410, 469.

———. "Springs for Chain Cables." *Nautical Magazine* 38 (1869): 338.

Walker, J. M. "The Meteorological Societies of London." *Weather* 48.11 (November 1993): 364–72. Available online at www.rmets.org/pdf/metsoclondon.pdf.

Walton, Oliver C. "Officers or Engineers? The Integration and Status of Engineers in the Royal Navy, 1847–60." *Historical Research* 77.196 (May 2004): 178–201.

Whalen, James M. "The Great Saxby Gale of October 1869." *The Beaver* 75.5 (Oct/ Nov 1995): 40–44.

Image Credits

American Journal of Science and Arts: 44 (1857), 45 (1831)
Author: 171, 207
Belcher's Farmer's Almanack, 1896: 36
British Library: 33
Charlie O'Reilly: 208
Chris Fogerty, Canadian Hurricane Centre: 142, 148
David T. Hughes, *Sheerness Naval Dockyard & Garrison*, 2002: 91
Donald Alward and the Albert County Museum, NB: 113
Elly Desplanque: 234
Fourth Meteorological Report, U.S. Senate, Ex. Doc. 65, 34th
    Congress, 3rd session, 42: 54
Geological Survey of Canada: 231, 232
H. E. L. Mellersh, *FitzRoy of the Beagle*, 1968: 76
Library of Congress: 61
NOAA: 222
Nova Scotia Department of Agriculture and Marketing: 241
Pennsylvania Academy of Fine Arts: 29
Picturesque Canada, Vol. 2, 1882: 175, 178, 180, 188
Provincial Archives of New Brunswick: 160
Punch magazine, September 10, 1881: 212
Saint John Telegraph and Morning Journal, 1869: 203
Stephen Saxby, *Saxby's Weather System*, 1864: 103
Scribner's magazine: 98 (February 1871), 136 (March 1871), 154
    (December 1870)
W. H. Davenport Adams, *The Garden Isle: The History,
    Topography, and Antiquities of the Isle of Wight*, 1856: 51

Index

Abbe, Cleveland, 10, 14, 128–29, 214, 216, 224
Airy, Sir George, 10, 108, 110, 129
Allison, Frederick, 10, 124–27, 195–97, 215–16
Allnatt, R. H., 10, 129, 206–207
almanacs, 28–34, 34–37, 38–39, 102, 195, 219
American Storm Controversy, 13, 56, 67, 85
apogee (moon), 102, *103*, 185, 226, 263
Arago, François, 10, 58, 73, 85–86, 104
astrology
  weather predictions based on, 32–33, 40–41
  influences on the body, 38
"astro-meteorology," 32–33, 40–41
astronomical tide, 230, 262
Atlantic Multi-decadal Oscillation (AMO), 235–36, 237, 262

Bache, Alexander Dallas, 10, 76–77
Bay of Fundy, 182–86, *188*, 225–27, 231–35, 238, 240
Belcher, Clement Horton, 10, 35
Belcher's Farmer's Almanac, 35, 195
British Almanac, 31
Browne, Walter Lord, 213
Bryson's Canadian Farmer's Almanac, 35, 39
Buys Ballot, Christoph, 10, 86
Buys Ballot's law, 86

Canadian Almanac, and Repository of Useful Knowledge, 34–35
Canadian Hurricane Centre, 217–18
chimney effect, 53
Civil War (U.S.), 14, 97–99, 105–106, 116
climate change, 236–37, 238

clouds, 39, 53–54
Coriolis effect, 13, 59–60, 84, 137, 138, 262
Coriolis, Gaspard-Gustave de, 10, 13, 59
Crimean War, 78–79, 85
Cunnabell, William, 10, 35, 37
Cunnabell's Nova Scotia Almanac and Farmer's Manual, 35, *36*, 37
cyclones. *See* tropical cyclones

Dampier, William, 44
Dawson, William Bell, 225–27, 230
Drum Stations, 215–16

Earle, James E., 174–76, 198–99
electricity, as cause of storms, 48, 58, 69, 102
El Niño, 236, 262
Espy, James Pollard, 10, 13, 48, 53–59, 65, 85

Farmer's Almanac (U.S.), 34
Farmer's Almanac (UK), 39
Farrer, T. H., 97, 118
Fenwick, William, 70, 81, 90
Ferrel, William, 10, 13, 83–85
Fitton, William H., 51
FitzRoy, Robert, 11, 13, 14, 73–75, *76*, 78, 79, 94, *98*, 100, 116–17
  criticism of, 117–18
  first weather forecasts, 14, 109–11
  meeting with Saxby, 107, 113–14
  storm warning system, 14, 95–97, *98*, 100–101, 106
flood prevention, 240–41
forecasting
  origin of term, 73, 110
  first weather forecasts, 109, 128, 214
  public response to, 110–11, 117–18
Franklin, Benjamin, 11, 13, 26–28, *29*, 34

frontal storms, 137, 262

Galton, Francis, 118, 119
Ganong, William, 224
Genii, 170–73, 200
Glaisher, James, 11, 31
global warming. *See* climate change
Great New England Hurricane (1938), 221
Great Norfolk and Long Island Hurricane, 43

Hare, Robert, 11, 58
Henry, Joseph, 11, 13, 49, 66–67, 75, 97, 116, 118–19, 126
Herschel, Sir John, 56, 58, 114, 115
Herschel, Sir William, 39, 114
Howard, Luke, 11, 39
HURDAT, 228–29
hurricanes, 27, 41, 43–44, 45–46, 59–60, 173–74, *222*, 228, 262
    Atlantic, 15, 47, 137, 146–47, 216–17, 220–21, 236
    formation of, 54–55, 136–40
    Hazel, 221
    Juan, 15–16
    wind speeds, 47–48, 140

Illustrated London Almanack, 30, 31

Kingston, George Templeman, 11, 125–26, 215–16
Knight, Charles, 11, 31

La Niña, 236, 263
latent heat, 11, 13, 54, 263
Leighton, George Cargill, 11, 31
Le Verrier, Urbain, 11, 85–86
lunar theory, 17, 38–39, 40–41, 100–108, *103*, 111–15, 213
    public reception of, 104, 105, 107, 112, 114–15

Maury, Matthew Fontaine, 11, 13, *61*, 61–63, 75–77, 79, 83, 98–99, 110
    Wind and Current Charts, 63, 83
    The Physical Geography of the Sea, 83, 84
    The Weather Book, 111
Meeley, James, 169–70, 171–72
Meteorological Society of London, 33–34.
    See also Royal Meteorological Society
moon
    apogee, 102, *103*, 185, 226, 263

declination, 101
perigee, 102, *103*, 185, 226, 249, 263
proposed influences of, 37–38
proposed influence on weather, 38–39, 40–41
    See also lunar theory
Moore's. See *Old Moore's Almanack*
Morrison, Richard James, 11, 32–34, *33*, 104, 118, 206
Morse, Samuel, 11, 13
Murphy, Patrick, 12, 31–32
Murphy's Weather Almanac, 31

national meteorological service (Canada), 125–26
naval instruction, history of, 87–89
neap tides, 185, 226, 263
Nepheloscope, 53, *54*
New Brunswick, 157–58, 165–69, *171*, 173–81, *175*, *178*, *180*, 189–91, 214–17
    aftermath of Saxby's storm in, 198–204
    other tropicals storms in, 220, 222–24
    See also Bay of Fundy
Newfoundland, 15, 221
New York, 158–62
Nova Scotia, 15, 124–25, 187, 191–97, 205–206, 221, 238–43

Old Moore's Almanack, 29, 30
Olmsted, Denison, 12, 45, 57

Partridge's (almanac), 30
perigee (moon), 102, *103*, 185, 226, 249, 263
Piddington, Henry, 12, 47, 48
Poor Richard's Almanack, 34
Poor Robin (almanac), 30–31
portents (weather), 39–40
post-tropical transition, 147, 263

Redfield, William C., 12, 13, 41–47, *44*, 48, 55–59, 63, 65, 84–85
Reid, William, 12, 46–47
Riccioli, Giovanni Battista, 60
Ring, Ken, 213
Royal Charter, 94, 103
Royal Meteorological Society, 34
Russell, John Scott, 80

Sabine, Edward, 80
Saffir-Simpson scale, 163, 220, 246–47

Saxby's Gale, 15–16
  formation of, 137–41, 147–48
  in the U.S., 141–46, 148, 149–50, 151–
    56, 163–65, 205
  in New Brunswick, 165–69, 173–81
  in Nova Scotia, 187, 191–97
  prediction of, 121–22, 126–27
  weather during, 248–49
  wind speeds, 138, 149, 163, 229
Saxby, Gavin Frank, 123–24
Saxby, Stephen Martin, 12, 14, 16, 17–18,
    113
  as a child, 25–26
  early life and career, 49–52, 63–65,
    68–70, 72, 79–80, 82
  inventions, 81–82
  as Naval Instructor at Sheerness, 88–92,
    119–20, 122–24
  lunar theory, 17, 40–41, 100–108, 103,
    111–15, 121, 210, 212
  prediction of Saxby's Gale, 121–22,
    129–31, 149, 205–206
  after the storm, 205–13
Scoresby, William, 80
sea level change, 237–38
Shipwrecked Mariners' Benevolent
    Society, 68, 105
shipwrecks, 95, 149–51, 162, 163, 165, 166,
    167. See also Genii; Royal Charter;
    Tayleur
Smith, Augustus, 111
St. Andrews, NB, 165–66
standardized weather observations, 75–77
steam-powered vessels, 42–43, 86–87. See
    also Thames
storm surge, 175–76, 188, 230, 263
storm warning systems, 86, 96–97, 100,
    106, 214–17

Tayleur, 71, 80, 81
telegraph, role of, 11, 13, 65–67, 73, 79, 86,
    96, 109, 127, 158
Thames, 24–25
Thomas, Robert B., 34
tidal range, 182–83, 226, 263
tides, 84, 182–85
  neap, 185, 226, 263
  spring, 185, 230, 263
Toaldo, Giuseppe, 104
troughs, 137, 264
tropical depressions, 137, 263
tropical storms, 139, 263

tropical waves, 137, 264
typhoons, 44, 47, 264

Village Belle, 135–36
Vines, Father Benito, 216–17

"weather prediction," 30, 31, 35, 37, 39
Winchester, E. B., 158–62, 200–201
wind shear, 236, 264

Zadkiel Tao-Sze. See Morrison, Richard
    James.